项目开发实战

（微视频版）

陈强◎编著

清华大学出版社
北京

内 容 简 介

Java 语言是当今使用最广泛的开发语言之一，在开发领域中占据重要的地位。本书通过 8 个综合项目的实现过程，详细讲解了 Java 语言在实践项目中的综合运用。第 1 章讲解了门户网站用户大数据分析系统的具体实现流程；第 2 章讲解了微信商城系统的具体实现流程；第 3 章讲解了图书借阅管理系统的具体实现流程；第 4 章讲解了物业管理系统的具体实现流程；第 5 章讲解了仿《羊了个羊》游戏的具体实现流程；第 6 章讲解了智能运动健身系统的具体实现流程；第 7 章讲解了图书市场数据分析系统的具体实现流程；第 8 章讲解了基于深度学习的音乐推荐系统的具体实现流程。在具体讲解每个项目时，都遵循项目的流程来讲解，从接到项目到具体开发，直到最后的调试和发布的过程，讲解循序渐进，穿插讲解了这样做的原因，深入讲解每个重点内容的具体细节，引领读者全面掌握 Java 语言。

本书不但适用于 Java 语言的初学者，也适用于有一定 Java 语言基础的读者，同时还可以供有一定经验的程序员参考。

本书封面贴有清华大学出版社防伪标签，无标签者不得销售。
版权所有，侵权必究。举报: 010-62782989, beiqinquan@tup.tsinghua.edu.cn。

图书在版编目(CIP)数据

Java 项目开发实战：微视频版/陈强编著. —北京：清华大学出版社，2024.5
ISBN 978-7-302-65987-7

Ⅰ. ①J… Ⅱ. ①陈… Ⅲ. ①JAVA 语言—程序设计 Ⅳ. ①TP312.8

中国国家版本馆 CIP 数据核字(2024)第 068305 号

责任编辑：魏　莹
封面设计：李　坤
责任校对：李玉茹
责任印制：杨　艳

出版发行：清华大学出版社
网　　址：https://www.tup.com.cn, https://www.wqxuetang.com
地　　址：北京清华大学学研大厦 A 座　　邮　编：100084
社 总 机：010-83470000　　邮　购：010-62786544
投稿与读者服务：010-62776969, c-service@tup.tsinghua.edu.cn
质量反馈：010-62772015, zhiliang@tup.tsinghua.edu.cn

印 装 者：三河市科茂嘉荣印务有限公司
经　　销：全国新华书店
开　　本：185mm×230mm　　印　张：21.75　　字　数：434 千字
版　　次：2024 年 5 月第 1 版　　印　次：2024 年 5 月第 1 次印刷
定　　价：89.00 元

产品编号：102091-01

前　　言

项目实战的重要性

在竞争日益激烈的软件开发市场中，拥有良好的理论基础是非常重要的。然而，仅仅掌握理论知识是不够的，实践能力是将理论知识转化为实际应用的关键。它不仅能够帮助我们更好地理解和记忆所学的知识，还能够培养我们解决问题的能力和创新能力。

在计算机科学领域，项目实战是一个将理论知识转化为实际应用的宝贵机会。课堂教学和理论学习是基础，但只有通过实际项目的实践，才能真正掌握所学的知识，并将其运用到实际场景中。项目实战不仅提供了将理论知识应用于实际问题的平台，还能够培养解决问题的能力和创新思维能力。以下是项目实战的重要性及其带给个人的益处。

(1) 实践锻炼：通过参与项目实战，面临真实的编码挑战，可从中学习解决问题的能力和技巧。实践锻炼有助于个人逐渐熟悉编程语言、开发工具和常用框架，提高编码技术和代码质量。

(2) 综合能力培养：项目实战要求综合运用各个知识点和技术，从需求分析、设计到实现和测试等环节，可全方位培养您的综合能力。这种综合性的学习方式将使您成为一名全面发展的优秀程序员。

(3) 团队协作经验：项目实战通常需要与团队成员合作，这对培养团队协作和沟通能力至关重要。通过与他人合作，您将学会协调工作、共同解决问题，并加深对团队合作的理解和体验。

(4) 独立思考能力：项目实战要求我们在遇到问题时能够独立思考和找到解决方法。通过克服困难、完成挑战，培养自信和勇气，提高您独立思考和解决问题的能力。

(5) 实践经验加分：在未来求职过程中，项目实战经验将成为您的亮点。用人单位更看重具有实践经验的人，更倾向于选择那些能够快速适应工作环境并提供实际解决方案的人才。

为了帮助广大读者从一名编程初学者快速成长为有实践经验的开发高手，我们精心编写了本书。本书以实战项目为素材，从项目背景和规划开始讲解，一直到项目的调试运行和维护结束，完整展示了大型商业项目的运作和开发流程。

本书的特色

1) 以实践为导向

本书的核心理念是通过实际项目来让读者学习和掌握 Java 语言编程的方法和技巧。每个项目都很实用，涵盖了不同领域和应用场景，能帮助读者将所学的知识直接应用到实际项目中。

2) 项目新颖

本书中的 8 个实战项目贴合现实主流应用领域,项目新颖。书中的项目涉及了大数据分析、微信商城、热门游戏、智能运动健身、深度学习等内容,这些都是当今开发领域的热点。

3) 渐进式学习

本书按照难度逐渐增加的顺序组织内容,从简单到复杂,让读者能够循序渐进地学习和提高。每个项目都有清晰的目标和步骤,引导读者逐步实现相应的功能。

4) 选取综合性项目进行讲解

本书包含多个综合性项目,涉及不同的编程概念和技术。通过完成这些项目,读者能够综合运用所学的知识,培养解决问题的能力和提高系统设计的思维。

5) 提供解决方案和提示

每个项目都提供了详细的解决方案和提示,这些解决方案和提示旨在启发读者思考,并提供参考,帮助读者理解项目的实现细节和关键技术,但也鼓励读者根据自己的理解和创意进行探索和实现。

6) 实用的案例应用

本书的项目涉及多个实际应用领域,如游戏开发、数据管理、深度学习等。这些案例不仅有助于读者理解 Java 语言的应用,还能够培养读者解决实际问题的能力。

7) 结合图表,通俗易懂

本书案例给出了相应的程序和表格进行说明,以使读者领会其含义;对于复杂的程序,均结合程序流程图进行讲解,以方便读者理解程序的执行过程;在语言的叙述上,普遍采用了短句子、易于理解的语言,避免使用复杂句子和晦涩难懂的语言。

8) 给读者以最大实惠

本书的附配资源不仅有书中实例的源代码和 PPT 课件(读者可扫描右侧二维码获取),还有书中案例全程视频讲解,视频讲解读者可扫描书中二维码来获取。

扫码获取源代码

致谢

本书由陈强编著。在编写本书的过程中,我们始终本着科学、严谨的态度,力求精益求精,但疏漏之处在所难免,敬请广大读者批评、指正。最后,感谢清华大学出版社的编辑,是他们的严谨和专业才使得本书得以快速出版。

编 者

目　录

第 1 章　门户网站用户大数据分析系统 1

- 1.1　大数据介绍 2
  - 1.1.1　大数据的特征 2
  - 1.1.2　大数据技术的应用 2
- 1.2　系统设计 3
  - 1.2.1　背景介绍 3
  - 1.2.2　系统目标 3
  - 1.2.3　系统功能结构 4
- 1.3　数据库设计 4
- 1.4　爬虫请求分析 5
- 1.5　系统组织结构和运行流程图 ... 9
  - 1.5.1　系统组织结构 9
  - 1.5.2　系统运行流程图 9
- 1.6　实现核心模块 11
  - 1.6.1　HTTP 请求的执行 11
  - 1.6.2　数据库连接 14
  - 1.6.3　数据库 dao 操作 15
  - 1.6.4　实现相关实体类 19
- 1.7　数据爬取模块 21
  - 1.7.1　爬虫爬取初始化 21
  - 1.7.2　知乎网页下载 25
  - 1.7.3　解析知乎详情列表页 ... 30
- 1.8　代理功能模块 32
  - 1.8.1　代理功能模块初始化 ... 32
  - 1.8.2　代理初始化 34
  - 1.8.3　代理页下载线程池和代理测试线程池初始化 35
  - 1.8.4　代理爬取入口 37
  - 1.8.5　代理页面下载 38
  - 1.8.6　代理页面解析 40
  - 1.8.7　代理可用性检测 42
  - 1.8.8　代理序列化 44
- 1.9　数据可视化分析 44
  - 1.9.1　数据展示模块 45
  - 1.9.2　运行展示 47
- 1.10　项目开发难点分析 48

第 2 章　微信商城系统 49

- 2.1　微信商城系统介绍 50
- 2.2　系统需求分析 50
- 2.3　系统架构 51
  - 2.3.1　第三方开源库 51
  - 2.3.2　系统架构介绍 52
  - 2.3.3　开发技术栈 52
- 2.4　实现管理后台模块 53
  - 2.4.1　用户登录验证 53
  - 2.4.2　用户管理 56
  - 2.4.3　订单管理 58
  - 2.4.4　商品管理 62
- 2.5　实现小商城系统 67
  - 2.5.1　系统主页 67
  - 2.5.2　会员注册登录 69
  - 2.5.3　商品分类 76
  - 2.5.4　商品搜索 79
  - 2.5.5　商品团购 81
  - 2.5.6　购物车 87

2.6 本地测试 ... 89	3.7.2 忘记密码 ... 120
2.6.1 创建数据库 ... 89	3.7.3 新用户注册 ... 122
2.6.2 运行后台管理系统 ... 90	3.8 基本信息管理模块 ... 125
2.6.3 运行微信小商城子系统 ... 92	3.8.1 读者信息管理 ... 125
2.7 线上发布和部署 ... 94	3.8.2 图书信息管理 ... 132
2.7.1 微信登录配置 ... 94	3.8.3 借书处理模块 ... 136
2.7.2 微信支付配置 ... 95	3.8.4 还书处理模块 ... 137
2.7.3 配置邮件通知 ... 95	3.9 数据操作 ... 140
2.7.4 短信通知配置 ... 96	3.9.1 用户登录验证 ... 140
2.7.5 系统部署 ... 97	3.9.2 获取图书信息 ... 141
2.7.6 技术支持 ... 97	3.9.3 获取读者信息 ... 142
2.7.7 项目参考 ... 97	3.9.4 添加借阅记录信息 ... 143
第 3 章 图书借阅管理系统 ... 99	3.9.5 添加新书信息 ... 144
3.1 背景介绍 ... 100	第 4 章 物业管理系统 ... 147
3.2 系统分析 ... 100	4.1 背景介绍 ... 148
3.2.1 系统需求分析 ... 100	4.2 系统分析和设计 ... 148
3.2.2 系统功能分析 ... 101	4.2.1 系统需求分析 ... 148
3.3 数据库设计 ... 102	4.2.2 设计流程分析 ... 148
3.3.1 选择数据库 ... 102	4.2.3 系统模拟流程 ... 150
3.3.2 数据库结构的设计 ... 102	4.3 数据库设计 ... 150
3.4 系统框架设计 ... 105	4.3.1 选择数据库 ... 150
3.4.1 创建工程 ... 105	4.3.2 数据库结构设计 ... 151
3.4.2 导入引用包 ... 105	4.4 系统框架设计 ... 154
3.5 设计界面 ... 107	4.4.1 创建工程及设计主界面 ... 154
3.5.1 使用 JavaFX Scene Builder 设计界面 ... 107	4.4.2 数据库 ADO 访问类 ... 159
3.5.2 设计主界面 ... 108	4.4.3 系统登录模块设计 ... 161
3.6 为数据库表添加对应的类 ... 111	4.5 基本信息管理模块 ... 163
3.6.1 Book 类 ... 111	4.5.1 小区信息管理 ... 163
3.6.2 借阅类 Borrow ... 114	4.5.2 楼宇信息管理 ... 167
3.7 系统登录模块 ... 116	4.5.3 业主信息管理 ... 170
3.7.1 登录验证 ... 116	4.5.4 收费信息管理 ... 172
	4.5.5 查询单价清单 ... 175

4.6	消费指数管理模块	176
	4.6.1 业主消费录入	176
	4.6.2 物业消费录入	183
4.7	各项费用管理模块	183
	4.7.1 业主费用查询	184
	4.7.2 物业费用查询	186
4.8	系统测试	189

第 5 章 仿《羊了个羊》游戏 191

5.1	背景介绍	192
	5.1.1 游戏行业发展现状	192
	5.1.2 虚拟现实快速发展	192
	5.1.3 云游戏持续增长	193
	5.1.4 移动游戏重回增长轨迹	193
5.2	项目分析	194
	5.2.1 游戏介绍	194
	5.2.2 规划开发流程	194
	5.2.3 模块结构	195
5.3	准备工作	196
	5.3.1 创建工程	196
	5.3.2 准备素材	196
5.4	读取素材文件	197
5.5	组件模块	199
	5.5.1 实现方块类	199
	5.5.2 填充方块	200
	5.5.3 记录方块位置	201
	5.5.4 记录方块空间位置	205
5.6	容器模块	209
	5.6.1 游戏背景区	209
	5.6.2 卡槽	211
5.7	主程序	214
5.8	调试运行	217

第 6 章 智能运动健身系统 219

6.1	背景介绍	220
6.2	运动健身发展趋势	220
6.3	系统分析	221
	6.3.1 技术分析	221
	6.3.2 模块分析	222
6.4	系统主界面	222
	6.4.1 布局文件	222
	6.4.2 实现主 Activity	224
	6.4.3 系统服务	236
6.5	系统设置	244
	6.5.1 选项设置	244
	6.5.2 生成 GPX 和 KML 格式的文件	245
6.6	邮件分享提醒	248
	6.6.1 基本邮箱设置	248
	6.6.2 发送邮件	251
6.7	上传 OSM 地图	252
	6.7.1 授权提示布局文件	252
	6.7.2 文件上传	254
6.8	调试运行	256

第 7 章 图书市场数据分析系统 257

7.1	图书市场介绍	258
	7.1.1 图书市场现状分析	258
	7.1.2 图书市场背景分析	258
	7.1.3 图书市场发展趋势	259
7.2	系统分析	259
	7.2.1 系统介绍	260
	7.2.2 需求分析	260
7.3	系统模块和实现流程	260
7.4	爬虫抓取模块	261

- 7.4.1 网页概览 261
- 7.4.2 破解 JS API 反爬机制 264
- 7.4.3 爬虫抓取 Java 图书信息 268
- 7.4.4 爬虫抓取 Python 图书信息 272
- 7.4.5 爬虫抓取主分类图书信息类 274
- 7.4.6 爬虫抓取子分类图书信息类 280
- 7.5 大数据可视化分析 283
  - 7.5.1 搭建 Java Web 平台 284
  - 7.5.2 大数据分析并可视化计算机图书数据 286
  - 7.5.3 大数据分析并可视化近期 Java 书和 Python 书的数据 289
  - 7.5.4 大数据分析并可视化主分类图书数据 291
  - 7.5.5 大数据分析并可视化计算机子类图书数据 293

第 8 章 基于深度学习的音乐推荐系统 297

- 8.1 背景介绍 298
- 8.2 系统分析 298
  - 8.2.1 系统功能分析 298
  - 8.2.2 系统需求分析 298
  - 8.2.3 系统模块分析 299
- 8.3 系统架构分析 300
  - 8.3.1 MVC 架构 300
  - 8.3.2 深度学习 300
- 8.4 数据库设计 301
  - 8.4.1 数据库架构设计 301
  - 8.4.2 数据库结构设计 302
- 8.5 用户管理模块 305
  - 8.5.1 用户注册 305
  - 8.5.2 用户登录 309
  - 8.5.3 收藏歌曲 310
  - 8.5.4 用户评论和点赞 311
  - 8.5.5 音乐播放记录 315
  - 8.5.6 音乐下载 315
- 8.6 管理员管理模块 316
  - 8.6.1 信息搜索 316
  - 8.6.2 用户管理 319
  - 8.6.3 音乐管理 319
- 8.7 排行榜模块 323
  - 8.7.1 获取数据库数据 323
  - 8.7.2 展示排行榜数据 324
- 8.8 热门推荐模块 325
  - 8.8.1 Controller 文件 325
  - 8.8.2 获取数据库信息 325
- 8.9 个性化推荐模块 326
  - 8.9.1 展示个性化推荐信息 326
  - 8.9.2 实现 ServiceImpl 类 327
  - 8.9.3 随机梯度下降算法 330
  - 8.9.4 K 近邻分类算法 331
  - 8.9.5 协同过滤算法 332
  - 8.9.6 数据转换 334
- 8.10 项目测试 339

第1章 门户网站用户大数据分析系统

实施国家大数据战略是建立数字时代国家竞争优势的基础，世界各国都已充分认识到大数据对于国家的战略意义，并早早开始布局。本章将介绍使用 Java 语言开发一个门户网站用户大数据分析系统的方法，并详细介绍网络爬虫和数据可视化分析的流程。本章项目由网络爬虫+JSP+MySQL+Echarts 实现。

1.1 大数据介绍

大数据(big data)又称巨量资料,指的是所涉及的资料量规模巨大到无法通过目前主流软件工具,在合理时间内达到获取、管理、处理并整理成为帮助企业完成经营决策目的的数据。在20世纪80年代,"大数据"这个词就已经出现,但是它仅用来形容数据量大。随着计算机技术的不断发展,数据不再指简单的数字集合,而是指无法在有限时间内用传统的信息技术和软硬件工具对其进行感知、获取、管理、处理的方式。但对于"大数据"的具体定义,目前学术界尚未形成明确的认知。

扫码看视频

2012年,高德纳咨询公司认为:大数据是非常重要的信息资产,但它需要用新的运算方式来处理,以提高这项信息资产的决策力、洞察力,并用这些特征来描述大数据。麦肯锡(McKinsey)认为:想要在特定时间内对大数据的内容进行搜集、存储、分析、运用,依靠过去传统的数据处理方式已不能实现。

1.1.1 大数据的特征

关于"大数据"的特征描述,代表性的观点有:IBM将"大数据"的特点总结为3V,即大量化(volume)、多样化(variety)和快速化(velocity);著名的数据管理大师维克托•迈尔-舍恩伯格(Viktor Mayer-Schönberger)则认为大数据具有4个特点,即4V,在前面的基础上增加了value(价值密度低)。目前,4V特点已成为最基本的共识,这些特点使大数据有别于传统的数据概念。

数据量大是大数据的基本属性。想要收集大量数据是十分困难的,过去只有部分机构能够完成抽样调查,而现在,随着互联网的普及,用户通过智能化的媒介有意的分享或无意的点击、浏览都会产生大量数据。数据量大还体现在人们处理数据的方法和理念上。之前人们对事物的认知一直依据抽样调查,以部分数据来描述整体事物,但在某些领域,这种方法不能形成完整的描述,可能会忽略很多重要信息,甚至得到相反的结果。如今大多数领域的大数据依托云计算,不再采取部分样本来反映总体数据,提高了数据的准确性。

1.1.2 大数据技术的应用

数据的丰富意味着信息的丰富。海量信息的合理分析整合,对于企业管理层决策和政府部门决策都有很重要的指导意义。有实力的企业和政府部门都可以建立一套大数据处理系统来指导其作出决策。在数据大爆炸的时代,专门处理大数据的企业将迎来春天,因为还有很多企业不具备建立完善的大数据分析处理系统的能力。

随着大数据时代的到来，新的商业模式正在诞生，能否运用大数据技术完成商业模式的转型将是许多企业能否坚持下去的关键。同样，大数据时代的到来也给了新兴企业一个极佳的发展机遇。小米就是大数据时代新兴企业的一个典型代表，其依托互联网的营销模式和收集用户反馈信息进行分析处理以改善产品体验的方式，就是大数据技术的一个应用。

1.2 系统设计

在开发一个软件项目时，第一步永远是进行系统设计工作，预先规划整个项目的功能模块和运行流程。下面详细讲解本项目的系统设计过程。

扫码看视频

1.2.1 背景介绍

随着科学技术的飞速发展和社会经济水平的不断进步，互联网规模迅速膨胀。据中新社援引中国互联网络信息中心发布的第 51 次《中国互联网络发展状况统计报告》报道，截至 2022 年 12 月，中国网民规模达 10.67 亿，较 2021 年 12 月增长 3549 万，互联网普及率达 75.6%，这充分说明了互联网已经逐渐成为人类生活、学习所依赖的一部分。

网民每天的网络行为带来了网络用户行为数据的爆炸式增长。网络用户行为数据中蕴含着大量有价值、有意义的信息，通过对用户行为日志进行统计、分析，并将结果通过前台直观的报表展示，可以帮助营销商大致掌握用户的喜好，从中发现用户使用产品的规律。将这些规律与网站的营销策略、产品功能、运营策略相结合，对用户进行智能推荐，可以优化用户体验，实现更精细化和精准化的运营与营销，让产品销量获得更好的增长。此外，通过数据分析可以预测用户的行为倾向，为有关部门对网络舆论进行合理的监控和干预提供理论依据，还可以帮助公安部门针对犯罪嫌疑人进行网络行为监控等。

1.2.2 系统目标

根据系统需求分析，本项目的系统目标如下。
- 爬取知乎用户公开的个人资料信息。
- 构建专有爬虫 HTTP 代理池，突破同一客户端访问量的限制。
- 将数据保存到 MySQL 数据库。
- 多线程爬取，提高爬取速度。
- 对爬取到的知乎用户进行数据分析。

1.2.3 系统功能结构

根据系统需求，可以将系统分为两大模块：知乎爬取模块和代理功能模块。两大模块及其具体的功能模块如图 1-1 所示。

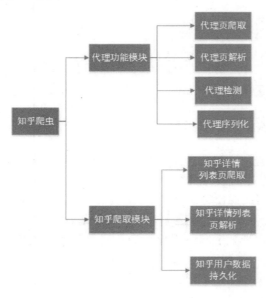

图 1-1 爬取系统功能模块划分

1.3 数据库设计

本系统用 MySQL 数据库存储数据，将爬取到的数据保存到 MySQL 数据库中。作为一个网络爬虫系统，相比其他管理类系统，数据库设计简单许多。本项目只涉及两个表：一个是 user 表，用于存放知乎用户信息；另一个是 url 表，用来存放 URL 数据。其中用户表 user 的设计结构如表 1-1 所示。

扫码看视频

表 1-1 user 表的设计结构

列 名	数据类型	字段说明
id	int(11)	自增 id
user_token	varchar(100)	个性地址令牌，唯一
location	varchar(100)	位置

续表

列　名	数据类型	字段说明
business	varchar(255)	行业
sex	varchar(255)	性别
employment	varchar(255)	企业
education	varchar(255)	教育
position	varchar(255)	职位
username	varchar(255)	用户名
url	varchar(255)	用户首页 url
agrees	int(11)	赞同数
thanks	int(11)	感谢数
asks	int(11)	提问数
answers	int(11)	回答问题数
posts	int(11)	文章数
followees	int(11)	关注数
followers	int(11)	粉丝数
hashId	varchar(255)	用户唯一标识

表 url 的具体设计结构如表 1-2 所示。

表 1-2　url 表的设计结构

列　名	数据类型	字段说明
id	int(11)	自增 id
user_md5_url	varchar(35)	url 爬取连接的 md5 摘要，唯一键

1.4　爬虫请求分析

就目前的情况，大部分网页上能看到的数据都是直接在网站后台生成好的(有的网页是在网站前端通过 JavaScripts 代码处理后显示的，例如数据混淆、加密等)，并直接在前台显示。虽然也有很多网站采用了 Ajax 异步加载功能，但归根结底它还是一个 HTTP 请求。只要能够分析出对应数据的请求来源，就能很容易地得到想要的数据。在接下来的内容中，将详细讲解 HTTP 请求的方法。

扫码看视频

(1) 以获取笔者知乎账户的所有关注用户资料为例。首先打开笔者关注列表页地址 https://www.zhihu.com/people/wo-yan-chen-mo/following，可以看到，主面板就是笔者关注用户列表。此时有 261 个关注用户，我们的目的就是获取这 261 个用户的个人资料信息。

(2) 打开 Chrome 浏览器，输入网址 https://www.zhihu.com/people/wo-yan-chen-mo/following，然后按 F12 键，依次单击 Network、XHR 按钮，勾选 Preserve log 和 Disable cache 两个复选框，如图 1-2 所示。

图 1-2　勾选 Preserve log 和 Disable cache 复选框

(3) 下拉滚动条，通过单击"下一页"按钮的方式来到第 4 页，获取对应请求(在翻页的过程中会有很多无关的请求，请不要理会)。待页面加载完成后，在请求列表中右击鼠标，在弹出的快捷菜单中选择 Save as HAR with content 命令。这个命令的功能是把当前请求(request)列表保存为 json 格式文本，然后使用 Chrome 浏览器打开这个文件，方法是单击浏览器中的"搜索"按钮(快捷键是 Ctrl+F)，在页面中提示输入搜索关键字，要注意这里中文采用了 Unicode 编码，此时直接搜索 9692(知乎账号"晨光文具"的关注者数)。这一步骤的目的是获取数据(关注用户的个人资料)的请求来源，具体如图 1-3 所示。

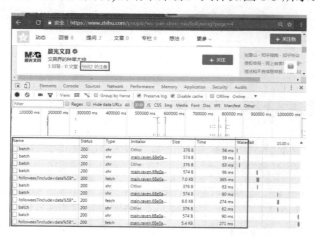

图 1-3　获取请求来源

(4) 由前面的搜索得出，关注用户的资料数据来自以下请求，如图 1-4 所示。URL 解码后为(命名为 url1)：

https://www.zhihu.com/api/v4/members/wo-yan-chen-mo/followees?include=data%5B*%5D.answer_count%2Carticles_count%2Cgender%2Cfollower_count%2Cis_followed%2Cis_following%2Cbadge%5B%3F(type%3Dbest_answerer)%5D.topics&offset=40&limit=20

图 1-4　请求 URL

从解码后的 URL 中可以看出，关注列表的数据并不是从 https://www.zhihu.com/people/wo-yan-chen-mo/following?page=4 同步加载而来的，而是直接通过 Ajax 异步请求 url1 来获得关注用户数据，然后通过 JavaScripts 代码填充数据。这里要注意用下方矩形圈住的 request 头 authorization，在实现代码的时候必须加上这个 Header(头)。这个数据并不是动态改变的，通过步骤(3)的方式可以发现它是来自一个 JavaScripts 文件。该步骤要注意的是，随着时间的推移，知乎可能会更新相关 API 接口的 URL，也就是说，通过步骤(3)得出的 URL 有可能并不是上面的 url1，但具体的分析方法是通用的。

(5) 经过多次测试分析后，可以得出以上 url1 的参数含义，如表 1-3 所示。

表 1-3　url1 的参数含义

参数名	类型	是否必填	值	说明
include	String	是	data[*]answer_count, articles_count	需要返回的字段(可以根据需要增加一些字段)

续表

参数名	类型	是否必填	值	说明
offset	int	是	0	偏移量(可以获取一个用户的所有关注用户资料)
limit	int	是	20	返回用户数(最大为20，超过20无效)

关于如何测试请求，建议采用以下三种方式。

- 原生 Chrome 浏览器：可以做一些简单的 GET 请求测试。这种方式有很大的局限性，不能编辑 HTTP Header(头)。如果直接(未登录知乎)通过浏览器访问 url1，会得到 401 错误的反馈代码。因为它没有带上 request 头 authorization，所以这种方式能测试一些简单且没有特殊 request 头的 GET 请求，如图 1-5 所示。

图 1-5　错误提示

- Chrome 插件 Postman：一个强大的 HTTP 请求测试工具，可以直接编辑 request，包括 cookies。它支持 GET、POST、PUT，几乎可以发送任意类型的 HTTP 请求。通过修改参数的值，来观察服务器响应数据的变化，从而确定参数的含义，如图 1-6 所示。

图 1-6　插件 Postman

❑ intellij idea ultimate 版自带的工具：打开方式是选择 Tools | Test RESTful Web Service 命令。也可以直接编辑 HTTP Header(包括 cookies)，然后发送请求，并且支持 GET、POST、PUT 等请求方式。

1.5 系统组织结构和运行流程图

在系统开发过程中，为了便于整个项目的管理和后期维护，需要规划好整个系统项目文件夹结构，并规划设计系统运行流程图。

扫码看视频

1.5.1 系统组织结构

按照系统功能划分文件夹，本项目的文件夹组织结构如图 1-7 所示。

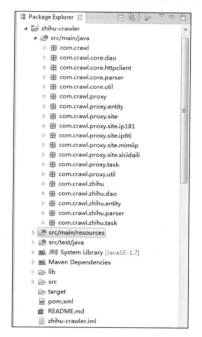

图 1-7 系统文件夹组织结构

1.5.2 系统运行流程图

整个项目程序的运行流程如图 1-8 所示。

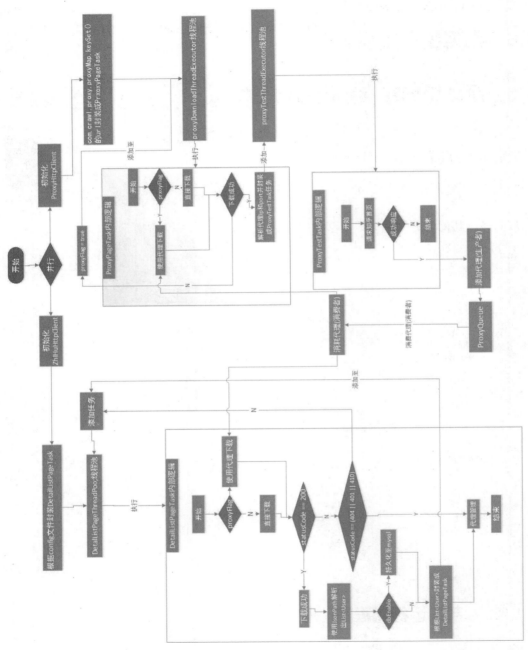

图 1-8 系统运行流程图

1.6 实现核心模块

为了提升本项目代码的可读性和重用性,本模块主要实现了整个项目的公共操作类、核心工具类、配置实体类,其中,公共操作包括数据库相关操作,核心工具包括 HTTP 请求工具。

扫码看视频

1.6.1 HTTP 请求的执行

本例是一个实战爬虫项目,最基本的功能便是实现 HTTP 请求。本功能的实现文件是 HttpClientUtil.java,本功能的难点在于 HttpClientContext 这个对象不是线程安全的,在多线程的情况下有一定的出错概率,并且问题不容易排查。官方文档对这点也有说明,建议读者在写代码的过程中要养成看官方文档的习惯,而不是遇到问题后才去搜索引擎寻找解决方法。网上相关资料很少,一般还是通过阅读源码才能解决这个问题。实例文件 HttpClientUtil.java 的主要代码如下。

```java
/**
 * 根据 url, 返回网页响应内容
 * @param url 网页地址
 * @return 网页内容
 * @throws IOexception
 */
public static String getWebPage(String url) throws IOException {
    //根据网页地址创建一个 httpget 请求
    HttpGet request = new HttpGet(url);
    //调用本类中的方法
    return getWebPage(request, "utf-8");
}

/**
 * 根据 request 请求对象, 返回响应内容
 * @param request http 请求
 * @return 网页内容
 * @throws IOException
 */
public static String getWebPage(HttpRequestBase request) throws IOException {
    //调用本类中的方法
    return getWebPage(request, "utf-8");
}

/**
```

```java
 * @param encoding 字符编码
 * @return 网页内容
 */
public static String getWebPage(HttpRequestBase request,
        String encoding) throws IOException {
    //调用本类中的方法，获取CloseableHttpResponse对象
    CloseableHttpResponse response = null;
    response = getResponse(request);
    //从response对象中获取响应状态码(200表示响应成功)，并输出到日志
    logger.info("status---" + response.getStatusLine().getStatusCode());
        //通过HttpClient工具类EntityUtils从response对象解析出网页响应内容
    String content = EntityUtils.toString(response.getEntity(),encoding);
    //释放http连接
    request.releaseConnection();
    //返回网页内容
    return content;
}

/**
 * 根据request对象，返回CloseableHttpResponse对象
 * @param request http请求
 * @return 网页内容
 * @throws IOException
 */
public static CloseableHttpResponse getResponse(HttpRequestBase request) throws
        IOException {
    //判断该请求是否有额外配置(如是否有代理情况、超时时间等配置)，如果没有则使用项目默认配置
    if (request.getConfig() == null){
        request.setConfig(requestConfig);
    }
    /**
     * 从Constants类中随机获取一个User-Agent，并设置到http request header，伪装成浏览器。
     * 防止服务器限制访问，此处随机的目的是防止服务器通过User-Agent来判断是否为同一用户
     *
     **/
    request.setHeader("User-Agent", Constants.userAgentArray[new
Random().nextInt(Constants.userAgentArray.length)]);
    /**
     * 创建HttpClientContext(HttpClient上下文，维护cookie相关)
     * 由于HttpClientContext不是线程安全，当有大量302状态码的http请求出现时，有很大概率会
       抛出异常
     * 所以此处将HttpClientContext设置为线程独享，共同维护同一个CookieStore对象
     **/
    HttpClientContext httpClientContext = HttpClientContext.create();
    // 设置Cookie
    httpClientContext.setCookieStore(cookieStore);
    // 携带http上下文执行http请求，并获得CloseableHttpResponse响应对象
```

```java
    CloseableHttpResponse response = httpClient.execute(request, httpClientContext);
    //返回 response 对象
    return response;
}
/**
 * 执行 http post 请求
 * @param postUrl 请求地址
 * @param params 请求参数，键值对
 * @return
 * @throws IOException
 */
public static String postRequest(String postUrl, Map<String, String> params) throws
        IOException {
    /**
     * 创建一个 httppost 请求对象
     */
    HttpPost post = new HttpPost(postUrl);
    //设置 post 请求参数
    setHttpPostParams(post, params);
    //根据 request 对象获取响应内容,响应编码为 utf-8
    return getWebPage(post, "utf-8");
}
/**
 * 设置 request 请求参数
 * @param request http post 对象
 * @param params post 请求参数
 */
public static void setHttpPostParams(HttpPost request,Map<String,String> params){
    //创建一个 NameValuePair http 表单键值对参数集合
    List<NameValuePair> formParams = new ArrayList<NameValuePair>();
    //遍历 map,根据其参数创建 NameValuePair 对象,并添加至 formParms list 集合中
    for (String key : params.keySet()) {
        formParams.add(new BasicNameValuePair(key,params.get(key)));
    }
    UrlEncodedFormEntity entity = null;
    try {
    //对参数进行 url 编码
        entity = new UrlEncodedFormEntity(formParams, "utf-8");
    } catch (UnsupportedEncodingException e) {
        e.printStackTrace();
    }
    //把表单参数添加至 request 请求对象
    request.setEntity(entity);
}
```

1.6.2 数据库连接

本系统提供了持久化知乎用户数据的功能,此功能涉及 JDBC 相关技术。本功能的实现文件是 ConnectionManager.java,其功能是创建数据库新连接、获取数据库连接和关闭连接。文件 ConnectionManager.java 的主要实现代码如下。

```java
/**
 * 创建一个新数据库 connection 并返回
 * @return 数据库 connection
 */
public static Connection createConnection(){
    //获取配置文件中的数据库 host 属性
    String host = Config.dbHost;
    //获取配置文件中的数据库登录用户名
    String user = Config.dbUsername;
    //获取配置文件中的数据库登录密码
    String password = Config.dbPassword;
    //获取配置文件中的数据库名
    String dbName = Config.dbName;
    //生成连接 url
    String url="jdbc:mysql://" + host + ":3306/" + dbName + "?characterEncoding=utf8";
    Connection con=null;
    try{
        //创建连接
        con = DriverManager.getConnection(url,user,password);
        logger.debug("success!");
    } catch(MySQLSyntaxErrorException e){
        logger.error("数据库不存在..请先手动创建数据库:" + dbName);
        e.printStackTrace();
    } catch(SQLException e2){
        logger.error("SQLException",e2);
    }
    return con;
}

/**
 * 返回当前 ConnectionManager 中的数据库连接,没有则创建新连接并返回
 * @return 数据库连接
 */
public static Connection getConnection(){
    try {
        if(conn == null || conn.isClosed()){
            conn = createConnection();
        } else{
            return conn;
```

```
        }
    } catch (SQLException e) {
        logger.error("SQLException",e);
    }
    return conn;
}

/**
 * 关闭数据库连接
 */
public static void close(){
    if(conn != null){
        try {
            conn.close();
        } catch (SQLException e) {
            logger.error("SQLException",e);
        }
    }
}
```

1.6.3 数据库 dao 操作

上一小节讲到了 connection 数据库连接功能，connection 连接主要是为 dao(data access object)层服务的，下面开始介绍该系统所有数据库相关操作(增、删、改、查)功能的实现。本功能的实现文件是 ZhiHuDaoImp.java，其功能是初始化数据库表。文件 ZhiHuDaoImp.java 的具体实现过程如下。

（1）初始化表功能，根据配置的数据库连接参数连接数据库，然后查询当前库是否已经创建表，如果不存在表则初始化 table。对应代码如下。

```
/**
 * 初始化 table
 */
public static void DBTablesInit() {
    ResultSet rs = null;
    //创建 properties 文件对象
    Properties p = new Properties();
    //获取一个数据库连接
    Connection cn = ConnectionManager.getConnection();
    try {
        //加载 config.properties 文件
        p.load(ZhiHuDaoImp.class.getResourceAsStream("/config.properties"));
        //查询 url table
        rs = cn.getMetaData().getTables(null, null, "url", null);
        //创建数据库 Statement 对象
```

```
        Statement st = cn.createStatement();
        if(!rs.next()){
            //不存在url表，创建url表
            st.execute(p.getProperty("createUrlTable"));
            logger.info("url表创建成功");
        }
        else{
            logger.info("url表已存在");
        }
        //查询user表
        rs = cn.getMetaData().getTables(null, null, "user", null);
        if(!rs.next()){
            //不存在user表，创建user表
            st.execute(p.getProperty("createUserTable"));
            logger.info("user表创建成功");
        }
        else{
            logger.info("user表已存在");
        }
        //关闭数据库结果集
        rs.close();
        //关闭数据库操作对象
        st.close();
        //关闭连接
        cn.close();
    } catch (SQLException e) {
        e.printStackTrace();
    } catch (IOException e) {
        e.printStackTrace();
    }
}
```

(2) 实现插入用户功能。此处需要先判断数据库中是否存在该用户，如果存在，就不执行 insert 操作并返回 false。如果不存在，则执行 insert 操作，然后返回 true。对应代码如下。

```
/**
 * insert user 资料至数据库
 * @param cn 数据库连接
 * @param u 用户对象
 * @return insert 操作结果
 */
@Override
public boolean insertUser(Connection cn, User u) {
    try {
        //判断数据库是否存在该用户，存在则直接返回false
        if (isExistUser(cn, u.getUserToken())){
            return false;
```

```java
        }
        //创建 insert sql 语句
        String column = "location,business,sex,employment,username,url,agrees,
                thanks,asks," +
                "answers,posts,followees,followers,hashId,education,user_token";
        String values = "?,?,?,?,?,?,?,?,?,?,?,?,?,?,?,?";
        String sql = "insert into user (" + column + ") values(" +values+")";
        PreparedStatement pstmt;
        //根据数据库连接创建 PreparedStatement 对象
        pstmt = cn.preparedStatement(sql);
        //设置用户所在位置
        pstmt.setString(1,u.getLocation());
        //设置用户所在行业
        pstmt.setString(2,u.getBusiness());
        //设置用户性别
        pstmt.setString(3,u.getSex());
        //设置用户所在企业
        pstmt.setString(4,u.getEmployment());
        //设置用户名
        pstmt.setString(5,u.getUsername());
        //知乎主页 url
        pstmt.setString(6,u.getUrl());
        //获得的赞同数
        pstmt.setInt(7,u.getAgrees());
        //获得的感谢数
        pstmt.setInt(8,u.getThanks());
        //获得的提问数
        pstmt.setInt(9,u.getAsks());
        //获得的回答问题数
        pstmt.setInt(10,u.getAnswers());
        //获得的文章数
        pstmt.setInt(11,u.getPosts());
        //获得的粉丝数
        pstmt.setInt(12,u.getFollowees());
        //获得的关注数
        pstmt.setInt(13,u.getFollowers());
        //hashId，用户唯一标识
        pstmt.setString(14,u.getHashId());
        //教育
        pstmt.setString(15,u.getEducation());
        //用户令牌，用户唯一标识
        pstmt.setString(16,u.getUserToken());
        pstmt.executeUpdate();
        pstmt.close();
        logger.info("插入数据库成功---" + u.getUsername());
    } catch (SQLException e) {
        e.printStackTrace();
```

```
        } finally {
        }
        return true;
}
```

(3) 实现查询功能。根据 userToken 查询数据库中是否已爬取过该用户，如果数据库中存在该用户就返回 true，否则返回 false。对应代码如下。

```
/**
 * 根据 userToken 查询数据库是否存在该用户
 * @param userToken 用户令牌，唯一标识
 * @return 查询结果
 */
@Override
public boolean isExistUser(String userToken) {
    //调用本类中的方法
    return isExistUser(ConnectionManager.getConnection(), userToken);
}

/**
 * 根据 userToken 查询数据库是否存在该用户
 * @param cn 数据库连接
 * @param userToken 用户令牌，唯一标识
 * @return
 */
@Override
public boolean isExistUser(Connection cn, String userToken) {
    //查询 sql
    String isContainSql = "select count(*) from user WHERE user_token='" + userToken + "'";
    try {
        if(isExistRecord(isContainSql)){
            return true;
        }
    } catch (SQLException e) {
        e.printStackTrace();
    }
    return false;
}

/**
 * 根据查询语句，判断是否存在查询结果
 * @param sql 查询语句
 * @return
 * @throws SQLException
 */
@Override
public boolean isExistRecord(String sql) throws SQLException{
```

```java
    //调用本类中的方法
    return isExistRecord(ConnectionManager.getConnection(), sql);
}

/**
 * 根据查询语句，判断是否存在查询结果
 * @param cn 数据库连接
 * @param sql 查询语句
 * @return
 * @throws SQLException
 */
@Override
public boolean isExistRecord(Connection cn, String sql) throws SQLException {
    int num = 0;
    PreparedStatement pstmt;
    //创建 PreparedStatement 对象
    pstmt = cn.prepareStatement(sql);
    //执行查询，并返回结果
    ResultSet rs = pstmt.executeQuery();
    while(rs.next()){
        num = rs.getInt("count(*)");
    }
    //关闭 ResultSet 对象
    rs.close();
    //关闭 PreparedStatement 对象
    pstmt.close();
    if(num == 0){
        return false;
    }else{
        return true;
    }
}
```

1.6.4 实现相关实体类

一般来说，实体类对应数据库表中的一条记录，也就是说，数据库表的一条记录是一个实体对象，而某行的一列就表示一个实体对象的一个属性。本项目实体类的实现文件是 User.java，具体代码如下。

```java
/**
 * 知乎用户资料
 */
public class User {
    //用户名
    private String username;
```

```java
//user token
private String userToken;
//位置
private String location;
//行业
private String business;
//性别
private String sex;
//企业
private String employment;
//企业职位
private String position;
//教育
private String education;
//用户首页url
private String url;
//赞同数
private int agrees;
//感谢数
private int thanks;
//提问数
private int asks;
//回答问题数
private int answers;
//文章数
private int posts;
//关注数
private int followees;
//粉丝数
private int followers;
// hashId, 用户唯一标识
private String hashId;

public String getUsername() {
    return username;
}

public void setUsername(String username) {
    this.username = username;
}
 //其他普通字段的get、set方法同上，此处省略部分get、set方法
@Override
public String toString() {
    return "User{" +
            "username='" + username + '\'' +
            ", userToken='" + userToken + '\'' +
            ", location='" + location + '\'' +
```

```
              ", business='" + business + '\'' +
              ", sex='" + sex + '\'' +
              ", employment='" + employment + '\'' +
              ", position='" + position + '\'' +
              ", education='" + education + '\'' +
              ", url='" + url + '\'' +
              ", agrees=" + agrees +
              ", thanks=" + thanks +
              ", asks=" + asks +
              ", answers=" + answers +
              ", posts=" + posts +
              ", followees=" + followees +
              ", followers=" + followers +
              ", hashId='" + hashId + '\'' +
              '}';
    }
}
```

1.7 数据爬取模块

数据爬取模块是整个项目的核心模块之一,主要包含的功能有整个爬虫的初始化、知乎页面的下载、知乎页面数据的解析、整个爬虫的运行流程控制、爬取异常处理、重试机制等。

扫码看视频

1.7.1 爬虫爬取初始化

1) authorization 字段的初始化

通过浏览器对知乎网站进行抓包后可知,抓取页面的请求不是普通的 GET 请求,而是要携带额外的 HTTP 验证头,即当通过爬虫方式来实现爬取的时候,需要每次携带该验证头才能请求成功。通过详细分析抓包可知,authorization 字段在 JavaScripts 脚本文件中,而 JavaScripts 的地址又需要通过知乎用户关注页面才能拿到,所以在初始化 authorization 字段时需要如下两个步骤。

(1) 请求并下载关注页面,解析出 authorization 所在 JavaScripts 文件的 URL。
(2) 请求下载该 URL 的数据,最终解析出 authorization 字段。

上述功能的实现文件是 ZhiHuHttpClient.java,对应代码如下。

```
/**
 * 初始化 authorization
 * @return
 */
```

```java
private String initAuthorization(){
    logger.info("初始化authorization中...");
    String content = null;
    //创建一个页面下载任务
    GeneralPageTask generalPageTask = new GeneralPageTask(Config.startURL, true);
    //执行下载任务，直接调用run方法
    generalPageTask.run();
    //获取下载成功的网页内容
    content = generalPageTask.getPage().getHtml();
    //创建一个正则表达式，获取authorization所在js文件地址的url
    Pattern pattern = Pattern.compile("https://static\\.zhihu\\.com/heifetz/"
                    + "main\\.app\\.([0-9]|[a-z])*\\.js");
    Matcher matcher = pattern.matcher(content);
    String jsSrc = null;
    if (matcher.find()){
        //解析出js文件的url
        jsSrc = matcher.group(0);
    } else {
        throw new RuntimeException("not find javascript url");
    }
    String jsContent = null;
    //创建一个页面下载任务，地址为刚刚解析出的js文件url
    GeneralPageTask jsPageTask = new GeneralPageTask(jsSrc, true);
    jsPageTask.run();
    //获取下载成功的js文件内容
    jsContent = jsPageTask.getPage().getHtml();
    //创建一个正则表达式，解析出authorization字段值
    pattern = Pattern.compile("oauth\\\"\\),h=\\\"(([0-9]|[a-z])*)\"");
    matcher = pattern.matcher(jsContent);
    if (matcher.find()){
        //获取authorization字段成功
        String authorization = matcher.group(1);
        logger.info("初始化authorization完成");
        return authorization;
    }
    throw new RuntimeException("not get authorization");
}
```

2) 列表详情页线程池初始化

本项目采用多线程的方式来提高爬取速度，也就是将每一个URL抽象成一个具体线程任务，而这个任务又在短时间内结束其生命周期。如果每执行一个任务就创建一个线程，当任务数量达到百万级别的时候，在线程创建上的开销是很大的，所以这里采用线程池模型来执行任务。线程池适用于执行那些任务多而耗时短的操作。它的一个基本原理就是：创建指定数量的线程后，当有新的任务到来时，直接从线程池中获取线程来执行任务，这样就减少了频繁创建线程的开销。线程池的主要初始化代码见 **ZhiHuHttpClient** 文件，对应

代码如下。

```java
/**
 * 初始化线程池
 */
private void intiThreadPool(){
    /**
     * 创建一个 corePoolSize 为 100, maxmumPoolSize 为 100, 任务队列长度为 2000 的线程池,
     * 用于执行知乎详情列表页下载解析任务, 其中 poolSize 可以通过配置文件修改。
     * 在线程池中, 当线程数量达到设定的最大值(maxmumPoolSize 为 100)且任务队列长度达到设定的
     * 最大值(2000)时, 如果此时有新的任务要添加到线程池中, 这些新任务将被直接丢弃, 而不会被执行。
     * 这里调用的是 SimpleThreadPoolExecutor, 继承 ThreadPoolExecutor, 可以通过构造方法直
     * 接为线程池命名。
     * 这里继承的目的是在输出日志中观察各个线程池的运行状态。
     */
    detailListPageThreadPool = new SimpleThreadPoolExecutor(Config.downloadThreadSize,
        Config.downloadThreadSize,
        0L, TimeUnit.MILLISECONDS,
        new LinkedBlockingQueue<Runnable>(2000),
        new ThreadPoolExecutor.DiscardPolicy(),
        "detailListPageThreadPool");
    //开启一个新线程, 用于监视列表详情页测试线程池执行情况
    new Thread(new ThreadPoolMonitor(detailListPageThreadPool,
"detailListPageThreadPool")).start();
}
```

3) 管理 ZhiHuHttpClient

整个项目启动后,不可能无休止地运行下去。当爬取指定数量的网页后,需要平滑地关闭整个爬虫功能,方法是轮询检测 detailListPageThreadPool 线程池的执行情况,当完成指定任务数后关闭该线程池,不再接受新的任务。此外,还需要关闭线程池监视工具类。当 detailListPageThreadPool 关闭后,关闭线程池连接和 ProxyHttpClient 类的 proxyTestThreadExecutor 线程池及 proxyDownloadThreadExecutor 线程池。本功能的实现文件是 ZhiHuHttpClient.java,对应代码如下。

```java
/**
 * 管理知乎 HttpClient
 * 关闭整个爬虫
 */
public void manageHttpClient(){
    //每秒执行一次轮询, 检测整个爬虫执行情况
    while (true) {
        /**
         * 下载网页数
         */
```

```
                long downloadPageCount = detailListPageThreadPool.getTaskCount();
                //下载网页数达到配置下载数,关闭detailListPageThreadPool 线程池
                if (downloadPageCount >= Config.downloadPageCount &&
                    !detailListPageThreadPool.isShutdown()) {
                    isStop = true;
                    //设置ThreadPoolMonitor,isStopMonitor 字段为true。关闭线程池监视类
                    ThreadPoolMonitor.isStopMonitor = true;
                    //关闭detailListPageThreadPool 线程池
                    detailListPageThreadPool.shutdown();
                }
                //判断detailListPageThreadPool 线程池是否关闭,完成关闭后,再关闭数据库连接
                if(detailListPageThreadPool.isTerminated()){
                    //关闭数据库连接
                    Map<Thread, Connection> map = DetailListPageTask.getConnectionMap();
                    for(Connection cn : map.values()){
                        try {
                            if (cn != null && !cn.isClosed()){
                                cn.close();
                            }
                        } catch (SQLException e) {
                            e.printStackTrace();
                        }
                    }
                    //关闭代理检测线程池
                    ProxyHttpClient.getInstance().getProxyTestThreadExecutor().shutdownNow();
                    //关闭代理下载页线程池
ProxyHttpClient.getInstance().getProxyDownloadThreadExecutor().shutdownNow();
                    break;
                }
                double costTime = (System.currentTimeMillis() - startTime) / 1000.0;//单位为s
                logger.debug("爬取速率: " + parseUserCount.get() / costTime + "个/s");
                try {
                    Thread.sleep(1000);
                } catch (InterruptedException e) {
                    e.printStackTrace();
                }
            }
        }
```

4) 爬取入口

通过配置 startUserToken,可以指定从哪一个知乎用户开始爬取。需要注意的是,爬取路线是顺着用户的关注用户一直往下爬取,所以配置的 startUserToken 必须有关注用户。方法是创建一个 DetailListPageTask,并添加至 detailListPageThreadPool 线程池中执行。该功能的实现文件是 ZhiHuHttpClient.java,对应代码如下。

```java
/**
 * 开始爬取
 */
@Override
public void startCrawl() {
    //调用本类中的方法，初始化 authorization 字段
    authorization = initAuthorization();
    //获取爬取入口用户 token
    String startToken = Config.startUserToken;
    //根据 token 构建请求 url
    String startUrl = String.format(Constants.USER_FOLLOWEES_URL, startToken, 0);
    //根据 url 创建一个 GET 请求
    HttpGet request = new HttpGet(startUrl);
    //设置 authorization header
    request.setHeader("authorization", "oauth " + ZhiHuHttpClient.getAuthorization());
    //创建一个 DetailListPageTask，并添加至 detailListPageThreadPool 中执行
    detailListPageThreadPool.execute(new DetailListPageTask(request, Config.isProxy));
    manageHttpClient();
}
```

1.7.2 知乎网页下载

1) 知乎 HTTP 请求抽象页任务实现

HTTP 请求是不能保证百分之百成功的，在爬取过程中需要处理各种可能的请求失败，以及是否使用代理、使用代理失败的后续处理逻辑、代理的耗时统计等。该功能的实现文件是 AbstractPageTask.java，对应代码如下。

```java
/**
 * 线程任务
 */
public void run(){
    long requestStartTime = 0l;
    HttpGet tempRequest = null;
    try {
        Page page = null;
        if(url != null){
            if (proxyFlag) {
                //使用代理
                tempRequest = new HttpGet(url);
                //从代理池延时队列中获取一个代理
                currentProxy = ProxyPool.proxyQueue.take();
                //判断代理是否为 Direct(直连)
                if(!(currentProxy instanceof Direct)){
                    //不是本机直接连接，创建一个 HttpHost 代理对象
```

```java
                    HttpHost proxy = new HttpHost(currentProxy.getIp(),
                            currentProxy.getPort());
                    //设置代理
                    tempRequest.setConfig(HttpClientUtil.
                        getRequestConfigBuilder().setProxy(proxy).build());
                }
                requestStartTime = System.currentTimeMillis();
                //执行 HttpGet 请求，获取响应内容
                page = zhiHuHttpClient.getWebPage(tempRequest);
            }else {
                //不使用代理
                requestStartTime = System.currentTimeMillis();
                page = zhiHuHttpClient.getWebPage(url);
            }
        } else if(request != null){
            if (proxyFlag){
                //使用代理，从代理池延时队列中获取一个代理
                currentProxy = ProxyPool.proxyQueue.take();
                //判断代理是否为 Direct(直连)
                if(!(currentProxy instanceof Direct)) {
                    //不是本机直接连接，创建一个 HttpHost 代理对象
                    HttpHost proxy = new HttpHost(currentProxy.getIp(),
                            currentProxy.getPort());
                    //设置代理
                    request.setConfig(HttpClientUtil.
                        getRequestConfigBuilder().setProxy(proxy).build());
                }
                requestStartTime = System.currentTimeMillis();
                //执行请求，获取响应内容
                page = zhiHuHttpClient.getWebPage(request);
            }else {
                //直接下载
                requestStartTime = System.currentTimeMillis();
                page = zhiHuHttpClient.getWebPage(request);
            }
        }
        long requestEndTime = System.currentTimeMillis();
        page.setProxy(currentProxy);
        //获取响应状态码
        int status = page.getStatusCode();
        //拼接日志
        String logStr = Thread.currentThread().getName() + " " + currentProxy +
                " executing request " + page.getUrl()  + " response statusCode:"
                + status + " request cost time:" +
                (requestEndTime - requestStartTime) + "ms";
        if(status == HttpStatus.SC_OK){
            /**
```

```java
             * 返回 SC_OK 状态不一定表示响应成功，由于部分异常代理，不会返回目标请求 url
             * 的内容。所以此处需要二次判断
             */
            if (page.getHtml().contains("zhihu")
                && !page.getHtml().contains("安全验证")){
                logger.debug(logStr);
                //代理请求次数+1
                currentProxy.setSuccessfulTimes
                        (currentProxy.getSuccessfulTimes() + 1);
                //记录代理总共请求耗时
                currentProxy.setSuccessfulTotalTime
                    (currentProxy.getSuccessfulTotalTime() +
                    (requestEndTime - requestStartTime));
                //计算成功请求平均耗时
                double aTime = (currentProxy.getSuccessfulTotalTime() + 0.0) /
                            currentProxy.getSuccessfulTimes();
                currentProxy.setSuccessfulAverageTime(aTime);
                currentProxy.setLastSuccessfulTime(System.currentTimeMillis());
                //处理响应成功网页，具体处理由子类实现
                handle(page);
            }else {
                /**
                 * 代理异常，没有正确返回目标 url
                 */
                logger.warn("proxy exception:" + currentProxy.toString());
            }
        }
        /**
         * 401--不能通过验证
         */
        else if(status == 404 || status == 401 ||
                status == 410){
            logger.warn(logStr);
        }
        else {
            logger.error(logStr);
            Thread.sleep(100);
            retry();
        }
    } catch (InterruptedException e) {
        logger.error("InterruptedException", e);
    } catch (IOException e) {
        //请求异常
   if(currentProxy != null){
        /**
         * 该代理可用，将该代理继续添加到 proxyQueue
         */
```

```
            currentProxy.setFailureTimes(currentProxy.getFailureTimes() + 1);
        }
        if(!zhiHuHttpClient.getDetailListPageThreadPool().isShutdown()){
            //重试,具体重试方法由子类实现
            retry();
        }
    } finally {
        if (request != null){
            //释放连接
            request.releaseConnection();
        }
        if (tempRequest != null){
            //释放连接
            tempRequest.releaseConnection();
        }
        if (currentProxy != null && !ProxyUtil.isDiscardProxy(currentProxy)){
            //代理过滤,失败次数达到一定条件,丢弃代理
            currentProxy.setTimeInterval(Constants.TIME_INTERVAL);
            ProxyPool.proxyQueue.add(currentProxy);
        }
    }
}
```

2) 知乎详情列表页任务功能

该功能是对下载成功的用户详情列表页进行后续处理,解析出用户资料并入库,包括对URL的去重处理、构造待爬取的DetailListPageTask任务。本功能的难点是DetailListPageTask执行任务时,又构造新的DetailListPageTask添加至线程池中。注意,当爬取一定数量的用户后,某一次任务爬取的用户全是已经爬取过的用户,这时就没有新的任务添加至线程池了,从而导致线程池一直处于空任务的状态。该功能的实现文件是DetailListPageTask.java,对应代码如下。

```
/**
 * 对下载成功的知乎用户列表详情页进行后续处理
 * @param page 网页
 */
@Override
void handle(Page page) {
    if(!page.getHtml().startsWith("{\"paging\"")){
        //代理异常,未能正确返回目标请求数据,丢弃
        currentProxy = null;
        return;
    }
    //从下载成功的详情列表页中解析出用户
    List<User> list = proxyUserListPageParser.parseListPage(page);
    for(User u : list){
```

```java
logger.info("解析用户成功:" + u.toString());
if(Config.dbEnable){
    //数据库可用，获取数据库connection
    Connection cn = getConnection();
    if (zhiHuDao.insertUser(cn, u)){
        //insert user
        parseUserCount.incrementAndGet();
    }
    //根据解析出的user信息，获取当前用户关注的用户数
    for (int j = 0; j < u.getFollowees() / 20; j++){
        if (zhiHuHttpClient.getDetailListPageThreadPool().getQueue().size()
                        > 1000){
            continue;
        }
        //构造获取当前用户所关注用户的url
        String nextUrl = String.format(USER_FOLLOWEES_URL, u.getUserToken(),
                        j * 20);
        /**
         * 在这里，我们对生成的URL进行MD5摘要计算，并将其插入数据库，以便进行去重操作。
         * 如果当前URL尚未被访问过，或者detailListPageThreadPool的activeCount
         *   为1时，我们执行以下操作。
         * 设置这个条件的目的是避免在爬取数量达到一定量后，在某个用户所关注的用户已经全
         *   部爬取完毕的情况下，防止线程池任务一直处于等待状态而导致爬取停止。
         */
        if (zhiHuDao.insertUrl(cn, Md5Util.Convert2Md5(nextUrl)) ||
                zhiHuHttpClient.getDetailListPageThreadPool().
                        getActiveCount() == 1){
            //根据生成的url构造HttpGet对象
            HttpGet request = new HttpGet(nextUrl);
            //设置authorization验证header
            request.setHeader("authorization", "oauth " +
                            ZhiHuHttpClient.getAuthorization());
            //创建DetailListPageTask，并添加至detailListPageThreadPool中
            zhiHuHttpClient.getDetailListPageThreadPool().execute
                            (new DetailListPageTask(request, true));
        }
    }
}
else if(!Config.dbEnable || zhiHuHttpClient.
        getDetailListPageThreadPool().getActiveCount() == 1){
    //不使用数据库，则不做去重处理
    parseUserCount.incrementAndGet();
    for (int j = 0; j < u.getFollowees() / 20; j++){
        //构造nextUrl
        String nextUrl = String.format(USER_FOLLOWEES_URL,
                        u.getUserToken(), j * 20);
        //根据url创建HttpGet
```

```
                    HttpGet request = new HttpGet(nextUrl);
                    //设置 authorization 验证 header
                    request.setHeader("authorization", "oauth " +
ZhiHuHttpClient.getAuthorization());
                    //创建 DetailListPageTask,并添加至 detailListPageThreadPool 中
                    zhiHuHttpClient.getDetailListPageThreadPool().execute(new
DetailListPageTask(request, true));
                }
            }
        }
    }
```

1.7.3 解析知乎详情列表页

本项目中的爬虫模块抓取的页面并不是普通的 HTML 标签文档,而是 json 格式的数据文档,所以使用 jsonPath 库来解析数据。在实现本功能时需要注意,为了兼容以前的代码,知乎服务器所返回的数据字段与本地 user 对象字段名大多不一样,所以这里采用反射的方式直接将值注入对象中。解析知乎详情列表页功能的实现文件是 ZhiHuUserListPageParser.java,对应代码如下。

```
/**
 * 根据网页对象,解析出用户资料列表
 * @param page
 * @return 用户资料列表
 */
@Override
public List<User> parseListPage(Page page) {
    List<User> userList = new ArrayList<>();
    String baseJsonPath = "$.data.length()";
    DocumentContext dc = JsonPath.parse(page.getHtml());
    Integer userCount = dc.read(baseJsonPath);
    for (int i = 0; i < userCount; i++){
        User user = new User();
        String userBaseJsonPath = "$.data[" + i + "]";
        //userToken
        setUserInfoByJsonPath(user, "userToken", dc, userBaseJsonPath +
                        ".url_token");
        //username
        setUserInfoByJsonPath(user, "username", dc, userBaseJsonPath + ".name");
        //hashId
        setUserInfoByJsonPath(user, "hashId", dc, userBaseJsonPath + ".id");
        //关注人数
        setUserInfoByJsonPath(user, "followees", dc, userBaseJsonPath +
                        ".following_count");
```

```java
        //位置
        setUserInfoByJsonPath(user, "location", dc, userBaseJsonPath +
                        ".locations[0].name");
        //行业
        setUserInfoByJsonPath(user, "business", dc, userBaseJsonPath +
                        ".business.name");
        //公司
        setUserInfoByJsonPath(user, "employment", dc, userBaseJsonPath +
                        ".employments[0].company.name");
        //职位
        setUserInfoByJsonPath(user, "position", dc, userBaseJsonPath +
                        ".employments[0].job.name");
        //教育
        setUserInfoByJsonPath(user, "education", dc, userBaseJsonPath +
                        ".educations[0].school.name");
        //回答数
        setUserInfoByJsonPath(user, "answers", dc, userBaseJsonPath +
                        ".answer_count");
        //提问数
        setUserInfoByJsonPath(user, "asks", dc, userBaseJsonPath + ".question_count");
        //文章数
        setUserInfoByJsonPath(user, "posts", dc, userBaseJsonPath +
                        ".articles_count");
        //粉丝数
        setUserInfoByJsonPath(user, "followers", dc, userBaseJsonPath +
                        ".follower_count");
        //赞同数
        setUserInfoByJsonPath(user, "agrees", dc, userBaseJsonPath +
                        ".voteup_count");
        //感谢数
        setUserInfoByJsonPath(user, "thanks", dc, userBaseJsonPath +
                        ".thanked_count");
        try {
            //性别
            Integer gender = dc.read(userBaseJsonPath + ".gender");
            if (gender != null && gender == 1){
                user.setSex("male");
            }
            else if(gender != null && gender == 0){
                user.setSex("female");
            }
        } catch (PathNotFoundException e){
            //没有该属性
        }
        userList.add(user);
}
return userList;
```

```
    }
    /**
     * jsonPath 获取值,并通过反射直接注入 user 中
     * @param user user 对象
     * @param fieldName user 对象中的字段名
     * @param dc 文档上下文
     * @param jsonPath jsonPath 表达式
     */
    private void setUserInfoByJsonPath(User user, String fieldName,
                                      DocumentContext dc , String jsonPath){
        try {
            //根据 jsonPath 表达式获取对应的值
            Object o = dc.read(jsonPath);
            //根据 field 字段名获取对象的 Field 对象
            Field field = user.getClass().getDeclaredField(fieldName);
            //设置为可被访问
            field.setAccessible(true);
            //设置 user 对象的 field 字段的值
            field.set(user, o);
        } catch (PathNotFoundException e1) {
            //no results
        } catch (Exception e){
            e.printStackTrace();
        }
    }
}
```

1.8 代理功能模块

代理功能模块是为知乎爬取模块服务的,突破同一客户端访问知乎服务器的并发连接限制,以此提高整个项目的爬取速度。代理功能模块主要包括的功能有代理页面的下载、代理页面的解析、代理的测试。代理打分丢弃是指在代理池中对每个代理进行评分,并根据得分来决定保留或丢弃该代理。

扫码看视频

1.8.1 代理功能模块初始化

为了提高代理的可重用性,同一个 Proxy 的请求速率会被限制,而该类正是通过 JDK 中的 DelayQueue(延时队列)来实现这一功能的。DelayQueue 的元素必须实现 Delayed 接口。在此需要注意的是,这个类中增加了一些额外的属性,用于统计代理的一些请求信息,便于对代理进行打分。定义 Proxy 代理类的实现文件是 Proxy.java,对应代码如下。

```java
/**
 * http 代理实体
 * 实现 Delayed 接口，作为 DelayQueue 的元素
 */
public class Proxy implements Delayed, Serializable{
    private static final long serialVersionUID = -7583883432417635332L;
    //使用该代理的最小间隔时间，单位为ms
    private long timeInterval;
    //代理ip地址
    private String ip;
    //代理端口
    private int port;
    //该代理是否可用
    private boolean availableFlag;
    //是否匿名
    private boolean anonymousFlag;
    //最近一次请求成功时间
    private long lastSuccessfulTime;
    //请求成功总耗时
    private long successfulTotalTime;
    //请求失败次数
    private int failureTimes;
    //请求成功次数
    private int successfulTimes;
    //请求成功平均耗时
    private double successfulAverageTime;
    public Proxy(String ip, int port, long timeInterval) {
        this.ip = ip;
        this.port = port;
        this.timeInterval = timeInterval;
        this.timeInterval = TimeUnit.NANOSECONDS.convert(timeInterval,
                    TimeUnit.MILLISECONDS) + System.nanoTime();
    }
    public String getIp() {
        return ip;
    }

    public void setIp(String ip) {
        this.ip = ip;
    }
    //其他普通字段的get、set方法同上，此处省略部分get、set方法

    @Override
    public int compareTo(Delayed o) {
        Proxy element = (Proxy)o;
        if (successfulAverageTime == 0.0d ||element.successfulAverageTime == 0.0d){
            return 0;
```

```java
            }
            return successfulAverageTime > element.successfulAverageTime ?
                    1:(successfulAverageTime < element.successfulAverageTime ? -1 : 0);
        }

        @Override
        public String toString() {
            return "Proxy{" +
                    "timeInterval=" + timeInterval +
                    ", ip='" + ip + '\'' +
                    ", port=" + port +
                    ", availableFlag=" + availableFlag +
                    ", anonymousFlag=" + anonymousFlag +
                    ", lastSuccessfulTime=" + lastSuccessfulTime +
                    ", successfulTotalTime=" + successfulTotalTime +
                    ", failureTimes=" + failureTimes +
                    ", successfulTimes=" + successfulTimes +
                    ", successfulAverageTime=" + successfulAverageTime +
                    '}';
        }
        //此处重写equals，如果ip地址和port相同，则表示是同一个代理
        @Override
        public boolean equals(Object o) {
            if (this == o) return true;
            if (o == null || getClass() != o.getClass()) return false;
            Proxy proxy = (Proxy) o;
            if (port != proxy.port) return false;
            return ip.equals(proxy.ip);

        }
        @Override
        public int hashCode() {
            int result = ip.hashCode();
            result = 31 * result + port;
            return result;
        }
        public String getProxyStr(){
            return ip + ":" + port;
        }
}
```

1.8.2 代理初始化

爬虫初始化获取有用代理是一个比较耗时的过程，经常会导致前期爬取速度非常慢。为了解决这个问题，需要让爬虫快速启动，并且能很快地爬取。在爬取数据过程中，每隔

一段时间把代理序列化至文件，然后每次启动文件再将代理反序列化至内存中。如果代理在最近一小时内使用过，则直接使用。在此需要注意的是，为什么选择一小时作为超时时间？因为目前网上公开的免费代理大多都有一个特点：使用的时效特别短，在公开一小时后还能使用的代理非常少。代理初始化功能的实现文件是 ProxyHttpClient.java，对应代码如下。

```java
/**
 * 初始化proxy
 */
private void initProxy(){
    Proxy[] proxyArray = null;
    try {
        //反序列化代理文件
        proxyArray = (Proxy[]) HttpClientUtil.deserializeObject(Config.proxyPath);
        int usableProxyCount = 0;
        for (Proxy p : proxyArray){
            if (p == null){
                continue;
            }
            //设置
            p.setTimeInterval(Constants.TIME_INTERVAL);
            p.setFailureTimes(0);
            p.setSuccessfulTimes(0);
            long nowTime = System.currentTimeMillis();
            if (nowTime - p.getLastSuccessfulTime() < 1000 * 60 *60){
                //上次成功离现在少于一小时
                ProxyPool.proxyQueue.add(p);
                ProxyPool.proxySet.add(p);
                usableProxyCount++;
            }
        }
        logger.info("反序列化proxy成功," + proxyArray.length + "个代理,可用代理"
                + usableProxyCount + "个");
    } catch (Exception e) {
        logger.warn("反序列化proxy失败");
    }
}
```

1.8.3 代理页下载线程池和代理测试线程池初始化

在本项目中，代理功能模块主要负责两种类型的任务。

第一种：根据 ProxyPool.java 文件中配置的 URL 去下载并解析代理网页任务。

第二种：检测解析出的 Proxy 的可用性。

本项目是通过创建 proxyTestThreadPool 和 proxyDownloadThreadPoll 两个线程池来分别执行这两种类型的任务。线程池的主要初始化代码见 ZhiHuHttpClient 文件，对应代码如下。

```java
/**
 * 初始化线程池
 */
private void initThreadPool(){
    //创建一个 corePoolSize 为 100,maxmumPoolSize 为 100,任务队列长度为 10000 的线程池,
    //用于执行代理测试任务
    proxyTestThreadExecutor = new SimpleThreadPoolExecutor(100, 100,
        0L, TimeUnit.MILLISECONDS,
        new LinkedBlockingQueue<Runnable>(10000),
        new ThreadPoolExecutor.DiscardPolicy(),
        "proxyTestThreadExecutor");
    //创建一个 corePoolSize 为 10, maximumPoolSize 为 10, 任务队列长度为
    //nteger.MAX_VALUE 的线程池,用于执行代理页面下载任务
    proxyDownloadThreadExecutor = new SimpleThreadPoolExecutor(10, 10,
        0L, TimeUnit.MILLISECONDS,
        new LinkedBlockingQueue<Runnable>(), "" +
        "proxyDownloadThreadExecutor");
    //开启一个新线程用于监视代理测试线程池执行情况
    new Thread(new ThreadPoolMonitor(proxyTestThreadExecutor,
            "ProxyTestThreadPool")).start();
    //开启一个新线程用于监视代理页下载线程池执行情况
    new Thread(new ThreadPoolMonitor(proxyDownloadThreadExecutor,
            "ProxyDownloadThreadExecutor")).start();
}
/**
 * 爬取代理
 */
public void startCrawl(){
    new Thread(new Runnable() {
        @Override
        public void run() {
            while (true){
                for (String url : ProxyPool.proxyMap.keySet()){
                    /**
                     * 本机首次直接下载代理页面
                     */
                    proxyDownloadThreadExecutor.execute(new ProxyPageTask(url, false));
                    try {
                        Thread.sleep(1000);
                    } catch (InterruptedException e) {
                        e.printStackTrace();
                    }
                }
                try {
```

```
                Thread.sleep(1000 * 60 * 60);
            } catch (InterruptedException e) {
                e.printStackTrace();
            }
        }
    }
    }).start();
    new Thread(new ProxySerializeTask()).start();
}
public ThreadPoolExecutor getProxyTestThreadExecutor() {
    return proxyTestThreadExecutor;
}

public ThreadPoolExecutor getProxyDownloadThreadExecutor() {
    return proxyDownloadThreadExecutor;
}
}
```

1.8.4　代理爬取入口

根据文件 ProxyPool.java 中配置的代理页 URL，逐一构造 ProxyPageTask，并将其添加至 proxyDownloadThreadPool 中，然后创建一个代理序列化任务。代理爬取入口功能的实现文件是 ProxyHttpClient.java，对应代码如下。

```
/**
 * 爬取代理
 */
public void startCrawl(){
    //开启一个新线程
    new Thread(new Runnable() {
        @Override
        public void run() {
            while (true){
                for (String url : ProxyPool.proxyMap.keySet()){
                    /**
                     * 首次本机直接下载代理页面
                     */
                    proxyDownloadThreadPool.execute(new ProxyPageTask(url, false));
                    try {
                        Thread.sleep(1000);
                    } catch (InterruptedException e) {
                        e.printStackTrace();
                    }
                }
                try {
```

```
                    //每隔1小时重新获取代理
                    Thread.sleep(1000 * 60 * 60);
                } catch (InterruptedException e) {
                    e.printStackTrace();
                }
            }
        }
    }).start();
    //创建代理序列化任务
    new Thread(new ProxySerializeTask()).start();
}
```

1.8.5 代理页面下载

根据文件 ProxyPool.java 中的代理页 URL 下载代理页面，下载成功后解析出代理。根据代理构造 ProxyTestTask，并添加至 proxyTestThreadPooll 中。下载失败的代理页面构造新的 ProxyPageTask，通过代理重新下载，直至下载成功。本功能的实现文件是 ProxyPageTask.java，对应代码如下。

```
public void run(){
    //获取当前时间戳，单位为ms
    long requestStartTime = System.currentTimeMillis();
    HttpGet tempRequest = null;
    try {
        Page page = null;
        if (proxyFlag){
            //使用代理下载，创建 HttpGet 请求对象
            tempRequest = new HttpGet(url);
            //从延时队列获取一个代理
            currentProxy = proxyQueue.take();
            //判断是否为本机直接连接
            if(!(currentProxy instanceof Direct)){
                //不是直接连接，创建一个 HttpHost 代理对象
                HttpHost proxy = new HttpHost(currentProxy.getIp(),
                                    currentProxy.getPort());
                //设置代理至创建的请求
                tempRequest.setConfig(HttpClientUtil.
                    getRequestConfigBuilder().setProxy(proxy).build());
            }
            //执行请求，获取网页内容
            page = proxyHttpClient.getWebPage(tempRequest);
        }else {
            //不使用代理，直接下载
            page = proxyHttpClient.getWebPage(url);
        }
```

```java
            page.setProxy(currentProxy);
            //获取响应状态码
            int status = page.getStatusCode();
            //获取当前时间戳,单位为ms,用于统计请求耗时
            long requestEndTime = System.currentTimeMillis();
            String logStr = Thread.currentThread().getName() + " "
                        + getProxyStr(currentProxy) + " executing request " +
                        page.getUrl()  + " response statusCode:" + status +
                    " request cost time:" + (requestEndTime - requestStartTime) + "ms";
            if(status == HttpStatus.SC_OK){
                //获取代理页成功
                logger.debug(logStr);
                handle(page);
            } else {
                //获取代理页失败
                logger.error(logStr);
                Thread.sleep(100);
                //重试
                retry();
            }
        } catch (InterruptedException e) {
            logger.error("InterruptedException", e);
        } catch (IOException e) {
            retry();
        } finally {
            if(currentProxy != null){
                currentProxy.setTimeInterval(Constants.TIME_INTERVAL);
                proxyQueue.add(currentProxy);
            }
            if (tempRequest != null){
                //释放连接
                tempRequest.releaseConnection();
            }
        }
    }
    public void retry(){
        //创建ProxyPageTask任务,通过代理下载
        proxyHttpClient.getProxyDownloadThreadPool().execute
                    (new ProxyPageTask(url, true));
    }
    /**
     * 处理下载成功的代理页
     * @param page
     */
    public void handle(Page page){
        if (page.getHtml() == null || page.getHtml().equals("")){
            return;
```

```
            }
            //根据url获取代理页面解析器
            ProxyListPageParser parser = ProxyListPageParserFactory.
                    getProxyListPageParser(ProxyPool.proxyMap.get(url));
            //解析出proxy list
            List<Proxy> proxyList = parser.parse(page.getHtml());
            for(Proxy p : proxyList){
                if(!ZhiHuHttpClient.getInstance().getDetailListPageThreadPool().
                    isTerminated()){
                    //获取ProxyPool读锁
                    ProxyPool.lock.readLock().lock();
                    //判断当前代理是否已被添加过
                    boolean containFlag = ProxyPool.proxySet.contains(p);
                    //释放ProxyPool读锁
                    ProxyPool.lock.readLock().unlock();
                    if (!containFlag){
                        //未被添加过，获取响应写锁，添加代理至proxySet
                        ProxyPool.lock.writeLock().lock();
                        ProxyPool.proxySet.add(p);
                        //释放写锁
                        ProxyPool.lock.writeLock().unlock();
                        //创建一个ProxyTestTask(代理测试任务)，并添加至proxyTest线程池
                        proxyHttpClient.getProxyTestThreadPool().execute
                            (new ProxyTestTask(p));
                    }
                }
            }
        }
    }
```

1.8.6 代理页面解析

代理页面和知乎详情列表的网页内容不太一致，目前所爬取的 4 个代理网页的内容是 HTML 标签文档，所以并没有采用 jsonPath 解析功能，而是采用 Jsoup 库进行解析，Jsoup 在解析 HTML 文档时非常灵活方便。代理页面解析功能的具体代码详见 com.crawl.proxy.site 包，实现过程如下所示。

(1) 登录代理网站 http://www.66ip.cn/，代理解析类的实现文件是 Ip66ProxyListPageParser.java，对应代码如下。

```
/**
 * http://www.66ip.cn/
 */
public class Ip66ProxyListPageParser implements ProxyListPageParser {
    /**
```

```
 * 根据 http://www.66ip.cn/网页内容解析出 proxy list
 * @param content 网页内容
 * @return proxy list
 */
@Override
public List<Proxy> parse(String content) {
    List<Proxy> proxyList = new ArrayList<>();
    if (content == null || content.equals("")){
        return proxyList;
    }
    //根据网页内容创建 Document 对象
    Document document = Jsoup.parse(content);
    //查找标签 table 的子标签 tr, 并且 tr 索引大于 1
    Elements elements = document.select("table tr:gt(1)");
    for (Element element : elements){
        //第一个 td 标签, text 为 ip
        String ip = element.select("td:eq(0)").first().text();
        //第二个 td 标签, text 为 port
        String port = element.select("td:eq(1)").first().text();
        //第三个 td 标签, text 为匿名标志
        String isAnonymous = element.select("td:eq(3)").first().text();
        if(!anonymousFlag || isAnonymous.contains("匿")){
            //只添加匿名代理至 proxyList 中
            proxyList.add(new Proxy(ip, Integer.valueOf(port), TIME_INTERVAL));
        }
    }
    return proxyList;
}
```

(2) 登录网站 http://www.ip181.com/,该代理解析类的实现文件是 Ip181ProxyListPageParser.java,对应代码如下。

```
/**
 * http://www.ip181.com/
 */
public class Ip181ProxyListPageParser implements ProxyListPageParser {
    /**
     * 根据 http://www.ip181.com/网页内容解析出 proxy list
     * @param content 网页内容
     * @return proxy list
     */
    @Override
    public List<Proxy> parse(String content) {
        //根据网页内容创建 Document 对象
        Document document = Jsoup.parse(content);
        //获取 table 标签下索引大于 0 的 tr 标签
```

```
        Elements elements = document.select("table tr:gt(0)");
        List<Proxy> proxyList = new ArrayList<>(elements.size());
        for (Element element : elements){
            //获取第一个td标签，text为ip
            String ip = element.select("td:eq(0)").first().text();
            //获取第二个td标签，text为port
            String port = element.select("td:eq(1)").first().text();
            //获取第三个td标签，text为匿名标志
            String isAnonymous = element.select("td:eq(2)").first().text();
            if(!anonymousFlag || isAnonymous.contains("匿")){
                //添加匿名代理至proxyList
                proxyList.add(new Proxy(ip, Integer.valueOf(port), TIME_INTERVAL));
            }
        }
        return proxyList;
    }
}
```

其他代理网站的代理原理同以上两个网站，本文不再提供具体实现方法。

1.8.7　代理可用性检测

从代理网站爬取代理，大部分代理是不可用的，需要对代理进行检测，可用的代理再拿来作为爬虫代理。检测的方式是通过访问知乎首页，看是否能返回正确的响应。该功能的实现文件是ProxyTestTask.java，对应代码如下。

```
/**
 * 代理检测task
 * 访问知乎首页，判断能否正确响应
 * 将可用代理添加到DelayQueue(延时队列)中
 */
public class ProxyTestTask implements Runnable{
    private final static Logger logger = LoggerFactory.getLogger(ProxyTestTask.class);
    private Proxy proxy;
    public ProxyTestTask(Proxy proxy){
        this.proxy = proxy;
    }

    /**
     * 多线程任务
     */
    @Override
    public void run() {
        //获取当前时间戳，单位为ms
```

```java
long startTime = System.currentTimeMillis();
//创建url为https://www.zhihu.com的HttpGet请求
HttpGet request = new HttpGet(Constants.INDEX_URL);
try {
    //配置request，设置超时时间和代理
    RequestConfig requestConfig =
            RequestConfig.custom().setSocketTimeout(Constants.TIMEOUT).
            setConnectTimeout(Constants.TIMEOUT).
            setConnectionRequestTimeout(Constants.TIMEOUT).
            setProxy(new HttpHost(proxy.getIp(), proxy.getPort())).
            setCookieSpec(CookieSpecs.STANDARD).
            build();
    request.setConfig(requestConfig);
    //执行请求，获取响应
    Page page = ZhiHuHttpClient.getInstance().getWebPage(request);
    //获取当前时间戳
    long endTime = System.currentTimeMillis();
    String logStr = Thread.currentThread().getName() + " " +
            proxy.getProxyStr() +
            " executing request " + page.getUrl() + " response statusCode:"
            + page.getStatusCode() +
            " request cost time:" + (endTime - startTime) + "ms";
    if (page == null || page.getStatusCode() != 200){
        //未能正确响应，直接返回
        logger.warn(logStr);
        return;
    }
    //释放连接
    request.releaseConnection();
    //记录日志
    logger.debug(proxy.toString() + "---------" + page.toString());
    logger.debug(proxy.toString() + "----------代理可用--------请求耗时:" +
            (endTime - startTime) + "ms");

    //正确响应，该代理可用，添加代理至延时队列
    ProxyPool.proxyQueue.add(proxy);
} catch (IOException e) {
    logger.debug("IOException:", e);
} finally {
    //延时队列
    if (request != null){
        request.releaseConnection();
    }
}
}
private String getProxyStr(){
```

```
            return proxy.getIp() + ":" + proxy.getPort();
    }
}
```

1.8.8 代理序列化

代理序列化的时间间隔为 1 分钟。把 ProxyPool 代理池延时队列中的代理序列化至磁盘文件中，代理序列化功能的实现文件是 ProxySerializeTask.java，对应代码如下。

```java
/**
 * 代理序列化 task
 */
public class ProxySerializeTask implements Runnable{
    private static Logger logger = LoggerFactory.getLogger(ProxySerializeTask.class);

    /**
     * 实现 Runnable 接口方法
     * 每隔 1 分钟序列化当前可用代理至文件
     */
    @Override
    public void run() {
        while (!ZhiHuHttpClient.isStop){
            try {
                Thread.sleep(1000 * 10 * 1);
            } catch (InterruptedException e) {
                e.printStackTrace();
            }
            Proxy[] proxyArray = null;
            //将 ProxyPool 中可用的代理添加至 proxyArray
            proxyArray = ProxyPool.proxyQueue.toArray(new Proxy[0]);
            //序列化代理至硬盘
            HttpClientUtil.serializeObject(proxyArray, Config.proxyPath);
            logger.info("成功序列化" + proxyArray.length + "个代理");
        }
    }
}
```

1.9 数据可视化分析

将爬取的知乎用户数据保存到数据库后，接下来开始分析用户的信息，并绘制出对应的可视化结果图。

扫码看视频

1.9.1 数据展示模块

在前文讲解了对数据的爬取过程，接下来可以对爬取的数据进行简单的分析和处理。具体的方式是用 SQL 查询出想要的数据，然后通过前端库 echarts 展示在浏览器中。该项目并不是一个 Web 项目，但是需要在浏览器中对数据进行可视化展示，所以需要通过 JDK 自带的轻量级 httpserver 作为 WebServer，后端采用 velocity 模板引擎对数据进行渲染。传递数据到浏览器端后，再通过 echarts 对数据做可视化渲染。

1. HTTP 请求处理

该功能负责对 httpserver 接收的请求进行处理，实现文件是 HttpRequestHandler.java，对应代码如下。

```java
/**
 * 具体请求处理方法
 * @param httpExchange
 * @throws IOException
 */
private void process(HttpExchange httpExchange) throws IOException
{httpExchange.sendResponseHeaders(200, 0);
    //获取 response 流
    OutputStream responseBody = httpExchange.getResponseBody();
    List<ChartVO> chartVOList = new ArrayList<>();
    //创建可视化 vo 对象
    chartVOList.add(new ChartVO("followers", "粉丝数", "知乎粉丝数 top10"));
    chartVOList.add(new ChartVO("agrees", "赞同数", "知乎赞同数 top10"));
    chartVOList.add(new ChartVO("thanks", "感谢数", "知乎感谢数 top10"));
    chartVOList.add(new ChartVO("asks", "提问数", "知乎提问数 top10"));
    chartVOList.add(new ChartVO("answers", "回答数", "知乎回答数 top10"));
    chartVOList.add(new ChartVO("posts", "文章数", "知乎文章数 top10"));
    chartVOList.add(new ChartVO("followers", "关注数", "知乎关注数 top10"));
    for(ChartVO vo : chartVOList){
        //调用本类中的方法
        gettop10(vo);
    }
    //调用本类中的方法，渲染数据
    String response = render(chartVOList);
    //write 数据至 response 流中
    responseBody.write(response.getBytes());
    responseBody.close();
}
```

2. 展示数据查询

该功能是通过调用 dao 模块实现的，需要从数据库中查询所需字段的 top10 数据，再根据返回的 vo 对象构造前端页面 echarts 数据字符串对象。该功能的实现文件是 HttpRequestHandler.java，对应的代码如下。

```java
/**
 * 根据 vo 对象，从数据库中查询对应数据 top10
 * 通过反射获取对应字段的值，然后构造前端所需要的格式数据
 * @param vo
 * @return
 */
private ChartVO getTop10(ChartVO vo){
    //创建数据库操作对象
    ZhiHuDao zhiHuDao = new ZhiHuDaoImp();
    //创建 sql 查询语句
    String followersTop10Sql = "select * from user order by " + vo.getColumnName()
                    + " desc limit 0, 10";
    //根据 sql 查询数据
    List<User> userList = zhiHuDao.queryUserList(followersTop10Sql);

    //创建前端页面 echarts 横坐标数组字符串对象
    StringBuilder xAxis = new StringBuilder();
    //创建前端页面 echarts 纵坐标数组字符串对象
    StringBuilder series = new StringBuilder();
    xAxis.append("[");
    series.append("[");

    //根据字段名构造对应字段 get 方法名
    StringBuilder methodName = new StringBuilder(vo.getColumnName());
    int c = methodName.charAt(0);
    c = c - 32;
    methodName.deleteCharAt(0);
    methodName.insert(0, (char) c);
    Method method = null;
    try {
        //获取字段对应的 method 对象
        method = User.class.getMethod("get" + methodName);
    } catch (NoSuchMethodException e) {
        e.printStackTrace();
    }

    //设置数据
    for (User user : userList){
        xAxis.append("'" + user.getUsername() + "',");
        try {
```

```
            int value = (int) method.invoke(user);
            series.append("'" + value + "',");
        } catch (IllegalAccessException e) {
            e.printStackTrace();
        } catch (InvocationTargetException e) {
            e.printStackTrace();
        }
    }
    xAxis.deleteCharAt(xAxis.length() - 1);
    series.deleteCharAt(series.length() - 1);
    xAxis.append("]");
    series.append("]");
    vo.setxAxis(xAxis.toString());
    vo.setSeries(series.toString());
    return vo;
}
```

1.9.2 运行展示

启动项目程序文件 Main.java，打开浏览器，输入网址 http://localhost:8080/zhihu-data-analysis 后可以看到爬取数据的实时统计情况。其中，top10 学校用户数统计如图 1-9 所示，知乎粉丝数 top10 用户统计如图 1-10 所示。

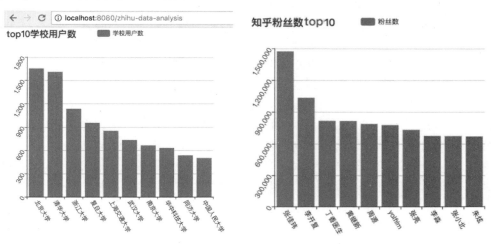

图 1-9　top10 学校用户数统计　　　　图 1-10　知乎粉丝数 top10 用户统计

知乎赞同数 top10 的用户统计如图 1-11 所示。

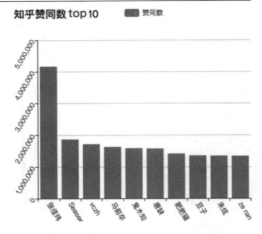

图 1-11　知乎赞同数 top10 的用户统计

1.10　项目开发难点分析

扫码看视频

　　现在来回顾整个系统的开发过程。对于简单的小型爬虫项目来说，当爬取数据量比较小的时候，项目没什么问题。但是当爬取的数据量比较大的时候，就会带来很多额外的问题，比如网站的统一客户端访问量的限制、数据的存储、URL 去重等。当前的知乎网站对同一客户端就有访问限制，而且一段时间内同一客户端请求量达到一定阈值时，会被知乎服务器禁止访问。此时若要爬取数据，就需要构建自己的免费爬虫代理池。具体的 HTTP 代理去哪儿找呢？可以用爬虫的方式去抓取网上免费公开的代理，然后把这些代理拿来爬取知乎上我们想要的用户数据。在此需要注意的是，网上免费公开代理的可用率非常低，几十个代理中可能只有一两个代理可用。如果想从这些代理中找出可用的代理，就需要编写专门的代理检测模块。拿到可用代理后，具体怎么使用这些代理呢？为了让这些代理达到最大利用率，在多线程抓取时，可通过延时队列来控制同一时刻同一代理最多只有一个请求在请求知乎服务器，并且使同一代理使用间隔为 1 秒，这样就能大大增加代理的复用性。

第 2 章

微信商城系统

本章将通过一个综合实例，讲解使用 Java 语言开发在线商城系统的过程。本系统的管理后端使用 Spring Boot 技术实现，管理前端使用 Vue 实现，购物商城前端通过微信小程序实现，购物商城后端通过 Spring Boot 实现。

2.1 微信商城系统介绍

在互联网大潮中,在线商城是指在网上建立一个在线销售平台,用户通过这个平台实现在线购买和提交订单,可以达到购买商品的目的。随着电子商务的蓬勃发展,在线商城系统在现实中得到迅猛发展。对于售方来说,可以节省店铺的经营成本;对于买方来说,可以实现即时购买,满足自己多方面的需求。

扫码看视频

随着移动智能手机的普及和微信用户的增多,过去曾经出现"移动互联网即将到来,微信是移动互联网入口"这一说。现在,微信用户高达 10 亿活跃量,确实证明了当初说法的正确性。

在过去的一段时间内,互联网 PC 端的电商市场被京东、淘宝、天猫等占领。商家要通过线上推广产品,只能花钱进驻平台,花钱做广告上首页,并在平台实现线上交易。商家想要开发属于自己的电商平台,没有大的资金、资源、实力几乎是不可能实现的。而互联网移动端仅限于开发 App、电子版网页,由于成本高、推广难度大等原因,导致很多中小企业放弃这方面的市场。这个时候,微信公众号商城是目前中小企业移动互联网转型的最佳渠道。与传统商城相比,微信公众号的优势如下。

(1)成本低,造价只有传统商城的四分之一。

(2)可以利用微信 10 亿用户群体,借助微信社交属性,完成裂变式推广,其覆盖的用户更广,还可以形成口碑式营销。

(3)可以实现 App 中 80%以上的功能。

在现在和将来的一段时间内,微信商城将是中小企业的必争之地。微商通过微信商城系统能够进行首页展示,以及产品库存、会员、分销、秒杀、拼团、会员奖励、财务、订单数据、物流配送、O2O 系统、分销、佣金分红等管理,有效帮助中小企业快速实现互联网转型。

2.2 系统需求分析

在现实应用中,一个典型的在线商城系统的构成模块如下。

1. 会员处理模块

为了方便用户购买商品,提高人气,系统设立了会员功能。成为系统会员

扫码看视频

后，用户可以对自己的资料进行管理，并且可以集中管理自己的订单。在线商城系统的用户处理模块必须具备以下功能。

(1) 会员注册：通过注册表单成为系统的会员，可以通过微信直接登录。

(2) 会员管理：会员不但可以管理个人基本信息，而且能够管理自己的订单信息。

2. 购物车处理模块

作为网上商城必不可少的环节，为满足用户的购物需求，系统设立了购物车功能。用户可以把需要的商品放到购物车中保存，提交在线订单后即可完成在线商品的购买。

3. 商品查寻模块

为了方便用户购买，系统设立了商品快速查寻模块，用户可以根据商品的信息快速找到自己需要的商品。

4. 订单处理模块

为方便商家处理用户的购买信息，系统设立了订单处理功能。通过该功能，可以实现用户购物车信息的及时处理，使用户尽快地收到自己的商品。

5. 商品分类模块

为了便于用户对系统商品的浏览，系统将商品划分为不同的类别，以便用户迅速找到自己需要的商品。

6. 商品管理模块

为方便系统的升级和维护，系统建立了专用的商品管理模块，实现商品的添加、删除和修改功能，以满足系统的更新需求。

2.3 系统架构

本系统是一个开源项目，在 GitHub 托管，名字为 litemall。开发团队一直在维护这个项目，具体的新功能和优化可登录 https://github.com/linlinjava/litemall 查看。本节将详细讲解该项目的具体架构。

扫码看视频

2.3.1 第三方开源库

为了避免重复工作，提高开发效率，本系统的架构是基于几款著名的第三方开源库，具体说明如下。

(1) nideshop-mini-program：基于 Node.js+MySQL 的开源微信小程序商城(微信小程序)。

(2) vue-element-admin：一个基于 Vue 和 Element 的后台集成方案。

(3) mall-admin-web：一个电商后台管理系统的前端项目，基于 Vue+Element 实现。

(4) biu：管理后台项目开发的脚手架，基于 vue-element-admin 和 Spring Boot 搭建，使用前后端分离方式开发和部署。

2.3.2 系统架构介绍

本系统是一个完整的前后端项目，包含 3 大部分 8 个模块，具体架构如图 2-1 所示。

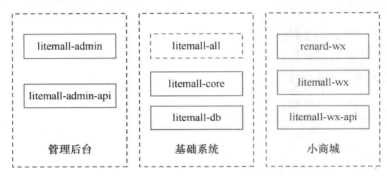

图 2-1 系统架构图

本系统各个模块的具体说明如下。

(1) 基础系统子系统(platform)：由数据库、litemall-core 模块、litemall-db 模块和 litemall-all 模块组成。

(2) 小商城子系统(wxmall)：由 litemall-wx-api 模块、litemall-wx 模块和 renard-wx 模块组成。

(3) 管理后台子系统(admin)：由 litemall-admin-api 模块和 litemall-admin 模块组成。

2.3.3 开发技术栈

在开发本项目各个模块的功能时，使用以下 3 种技术栈。

(1) Spring Boot 技术栈：采用 IntelliJ IDEA 开发工具，实现了 litemall-core、litemall-db、litemall-admin-api、litemall-wx-api 和 litemall-all 共计 5 个模块的功能。

(2) miniprogram(微信小程序)技术栈：采用微信小程序开发工具，分别实现了 litemall-wx 模块和 renard-wx 模块的功能。

(3) Vue 技术栈：采用 VSC 开发工具，实现了 litemall-admin 模块的功能。

2.4 实现管理后台模块

本项目的管理后台模块由前端和后端两部分实现，其中，前端实现模块是 litemall-admin，基于 Vue 技术实现；后端实现模块是 litemall-admin-api，基于 Spring Boot 技术实现。本节将详细讲解实现管理后台模块的具体过程。

扫码看视频

2.4.1 用户登录验证

(1) 在后台用户登录验证模块中，前端登录表单页面由文件 itemall\litemall-admin\src\views\login\index.vue 实现，此文件提供了一个简单的输入用户名和密码的表单，主要代码如下。

```
    <el-form ref="loginForm" :model="loginForm" :rules="loginRules" class="login-form"
        auto-complete="on" label-position="left">
      <div class="title-container">
        <h3 class="title">管理员登录</h3>
      </div>
      <el-form-item prop="username">
        <span class="svg-container svg-container_login">
          <svg-icon icon-class="user" />
        </span>
        <el-input v-model="loginForm.username" name="username" type="text" auto-
            complete="on" placeholder="username" />
      </el-form-item>

      <el-form-item prop="password">
        <span class="svg-container">
          <svg-icon icon-class="password" />
        </span>
        <el-input :type="passwordType" v-model="loginForm.password"
                  name="password"
            auto-complete="on" placeholder="password"
                      @keyup.enter.native="handleLogin" />
        <span class="show-pwd" @click="showPwd">
        <svg-icon icon-class="eye" />
          </span>
      </el-form-item>

      <el-button :loading="loading" type="primary"
                    style="width:100%;margin-bottom:30px;"
            @click.native.prevent="handleLogin">登录</el-button>
      <el-button :loading="loading" type="primary" style="width:100%;margin-bottom:30px;"
@click.native.prevent="handleLogin">登录</el-button>
```

```
        <div style="position:relative">
          <div class="tips">
            <span>超级管理员用户名：admin123</span>
            <span>超级管理员用户名：admin123</span>
          </div>
          <div class="tips">
            <span>商城管理员用户名：mall123</span>
            <span>商城管理员用户名：mall123</span>
          </div>
          <div class="tips">
            <span>推广管理员用户名：promotion123</span>
            <span>推广管理员用户名：promotion123</span>
          </div>
        </div>
      </el-form>
    </div>
  </template>
  <script>
  export default {
    name: 'Login',
    data() {
      const validateUsername = (rule, value, callback) => {
        if (validateUsername == null) {
          callback(new Error('请输入正确的管理员用户名'))
        } else {
          callback()
        }
      }
      const validatePassword = (rule, value, callback) => {
        if (value.length < 6) {
          callback(new Error('管理员密码长度应大于6'))
        } else {
          callback()
        }
      }
      return {
        loginForm: {
          username: 'admin123',
          password: 'admin123'
        },
        loginRules: {
          username: [{ required: true, trigger: 'blur', validator: validateUsername }],
          password: [{ required: true, trigger: 'blur', validator: validatePassword }]
        },
        passwordType: 'password',
        loading: false
      }
```

```
    },
    watch: {
      $route: {
        handler: function(route) {
          this.redirect = route.query && route.query.redirect
        },
        immediate: true
      }
    },
    methods: {
      showPwd() {
        if (this.passwordType === 'password') {
          this.passwordType = ''
        } else {
          this.passwordType = 'password'
        }
      },
      handleLogin() {
        this.$refs.loginForm.validate(valid => {
          if (valid && !this.loading) {
            this.loading = true
            this.$store.dispatch('LoginByUsername', this.loginForm).then(() => {
              this.loading = false
              this.$router.push({ path: this.redirect || '/' })
            }).catch(response => {
              this.$notify.error({
                title: '失败',
                message: response.data.errmsg
              })
              this.loading = false
            })
          } else {
            return false
          }
        })
      }
    }
  }
</script>
```

(2) 在后台用户登录验证模块中，后端登录验证功能通过视图文件 litemall-admin-api\src\main\java\org\linlinjava\litemall\admin\web\AdminCollectController.java 实现，主要功能是获取用户在表单中输入的信息，然后与数据库中存储的数据进行比较。文件 AdminCollectController.java 的主要代码如下：

```java
@PostMapping("/login")
public Object login(@RequestBody String body) {
    String username = JacksonUtil.parseString(body, "username");
    String password = JacksonUtil.parseString(body, "password");
    if (StringUtils.isEmpty(username) || StringUtils.isEmpty(password)) {
        return ResponseUtil.badArgument();
    }
    Subject currentUser = SecurityUtils.getSubject();
    try {
        currentUser.login(new UsernamePasswordToken(username, password));
    } catch (UnknownAccountException uae) {
        return ResponseUtil.fail(ADMIN_INVALID_ACCOUNT, "用户账号或密码不正确");
    } catch (LockedAccountException lae) {
        return ResponseUtil.fail(ADMIN_INVALID_ACCOUNT, "用户账号已锁定不可用");
    } catch (AuthenticationException ae) {
        return ResponseUtil.fail(ADMIN_INVALID_ACCOUNT, "认证失败");
    }
    return ResponseUtil.ok(currentUser.getSession().getId());
}
```

2.4.2 用户管理

在后台用户管理模块中，用户管理包含会员管理、收货地址、会员收藏、会员足迹、搜索历史和意见反馈 6 个子选项。接下来将简要讲解"会员管理"子选项功能的实现过程。其余选项的实现代码参见本书配套资源中的源码。

（1）在后台用户管理模块中，前端会员管理页面由文件 litemall\litemall-admin\src\views\user\user.vue 实现，在此文件顶部显示用户搜索表单和按钮，在下方分页列表显示系统内的所有会员信息。文件 user.vue 的主要代码如下。

```html
<!-- 查询和其他操作 -->
<div class="filter-container">
    <el-input v-model="listQuery.username" clearable class="filter-item" style=
        "width: 200px;" placeholder="请输入用户名"/>
    <el-input v-model="listQuery.mobile" clearable class="filter-item" style=
        "width: 200px;" placeholder="请输入手机号"/>
    <el-button class="filter-item" type="primary" icon="el-icon-search"
        @click="handleFilter">查找</el-button>
    <el-button :loading="downloadLoading" class="filter-item" type="primary"
        icon="el-icon-download" @click="handleDownload">导出</el-button>
</div>
<!-- 查询结果 -->
<el-table v-loading="listLoading" :data="list" size="small" element-loading-text=
    "正在查询中…" border fit highlight-current-row>
    <el-table-column align="center" width="100px" label="用户 ID" prop="id" sortable/>
```

```html
<el-table-column align="center" label="用户名" prop="username"/>
<el-table-column align="center" label="手机号码" prop="mobile"/>
<el-table-column align="center" label="性别" prop="gender">
  <template slot-scope="scope">
    <el-tag >{{ genderDic[scope.row.gender] }}</el-tag>
  </template>
</el-table-column>
<el-table-column align="center" label="生日" prop="birthday"/>

<el-table-column align="center" label="用户等级" prop="userLevel">
  <template slot-scope="scope">
    <el-tag >{{ levelDic[scope.row.userLevel] }}</el-tag>
  </template>
</el-table-column>
<el-table-column align="center" label="状态" prop="status">
  <template slot-scope="scope">
    <el-tag>{{ statusDic[scope.row.status] }}</el-tag>
  </template>
</el-table-column>
</el-table>
```

(2) 在后台用户管理模块中，后端会员管理功能通过视图文件 litemall-admin-api\src\main\java\org\linlinjava\litemall\admin\web\AdminUserController.java 实现，主要功能是获取系统数据库中的会员信息，然后将获取的信息列表显示在页面中；同时通过 total 计算数据库中会员的数量，然后根据这个数量进行分页显示。文件 AdminUserController.java 的主要代码如下。

```java
public class AdminUserController {
    private final Log logger = LogFactory.getLog(AdminUserController.class);
    @Autowired
    private LitemallUserService userService;

    @RequiresPermissions("admin:user:list")
    @RequiresPermissionsDesc(menu={"用户管理" , "会员管理"}, button="查询")
    @GetMapping("/list")
    public Object list(String username, String mobile,
                       @RequestParam(defaultValue = "1") Integer page,
                       @RequestParam(defaultValue = "10") Integer limit,
                       @Sort @RequestParam(defaultValue = "add_time") String sort,
                       @Order @RequestParam(defaultValue = "desc") String order) {
        List<LitemallUser> userList = userService.querySelective(username,
                mobile, page, limit, sort, order);
        long total = PageInfo.of(userList).getTotal();
        Map<String, Object> data = new HashMap<>();
        data.put("total", total);
        data.put("items", userList);
```

```
            return ResponseUtil.ok(data);
        }
}
```

2.4.3 订单管理

"订单管理"是后台商场管理的一个子选项,下面将详细讲解"订单管理"功能的实现过程。

(1) 在后台订单管理模块中,前端订单管理页面由文件 litemall\litemall-admin\src\views\mall\order.vue 实现,在此文件顶部显示订单搜索表单和按钮,在下方分页列表显示系统内的所有订单信息和订单详情按钮。文件 order.vue 的主要代码如下。

```
     <el-table-column align="center" min-width="100" label="订单编号" prop="orderSn"/>
     <el-table-column align="center" label="用户 ID" prop="userId"/>
     <el-table-column align="center" label="订单状态" prop="orderStatus">
       <template slot-scope="scope">
         <el-tag>{{ scope.row.orderStatus | orderStatusFilter }}</el-tag>
       </template>
     </el-table-column>
     <el-table-column align="center" label="订单金额" prop="orderPrice"/>
     <el-table-column align="center" label="支付金额" prop="actualPrice"/>
     <el-table-column align="center" label="支付时间" prop="payTime"/>
     <el-table-column align="center" label="物流单号" prop="shipSn"/>
     <el-table-column align="center" label="物流渠道" prop="shipChannel"/>
     <el-table-column align="center" label="操作" width="200" class-name="small-padding fixed-width">
       <template slot-scope="scope">
         <el-button v-permission="['GET /admin/order/detail']" type="primary" size="mini"
            @click="handleDetail(scope.row)">详情 </el-button>
         <el-button v-permission="['POST /admin/order/ship']"
v-if="scope.row.orderStatus==201"
            type="primary" size="mini" @click="handleShip(scope.row)">发货</el-button>
         <el-button v-permission="['POST /admin/order/refund']" v-if="scope.row.
            orderStatus==202" type="primary" size="mini" @click=
            "handleRefund(scope.row)">退款</el-button>
       </template>
     </el-table-column>
   </el-table>
   <pagination v-show="total>0" :total="total" :page.sync="listQuery.page" :
       limit.sync="listQuery.limit" @pagination="getList" />
   <!-- 订单详情对话框 -->
   <el-dialog :visible.sync="orderDialogVisible" title="订单详情" width="800">
       <el-form :data="orderDetail" label-position="left">
         <el-form-item label="订单编号">
```

```html
          <span>{{ orderDetail.order.orderSn }}</span>
        </el-form-item>
        <el-form-item label="订单状态">
          <template slot-scope="scope">
            <el-tag>{{ scope.order.orderStatus | orderStatusFilter }}</el-tag>
          </template>
        </el-form-item>
        <el-form-item label="订单用户">
          <span>{{ orderDetail.user.nickname }}</span>
        </el-form-item>
        <el-form-item label="用户留言">
          <span>{{ orderDetail.order.message }}</span>
        </el-form-item>
        <el-form-item label="收货信息">
          <span>(收货人){{ orderDetail.order.consignee }}</span>
          <span>(手机号){{ orderDetail.order.mobile }}</span>
          <span>(地址){{ orderDetail.order.address }}</span>
        </el-form-item>
        <el-form-item label="商品信息">
          <el-table :data="orderDetail.orderGoods" size="small" border fit
              highlight-current-row>
            <el-table-column align="center" label="商品名称" prop="goodsName" />
            <el-table-column align="center" label="商品编号" prop="goodsSn" />
            <el-table-column align="center" label="货品规格"
                prop="specifications" />
            <el-table-column align="center" label="货品价格" prop="price" />
            <el-table-column align="center" label="货品数量" prop="number" />
            <el-table-column align="center" label="货品图片" prop="picUrl">
              <template slot-scope="scope">
                <img :src="scope.row.picUrl" width="40">
              </template>
            </el-table-column>
          </el-table>
        </el-form-item>
        <el-form-item label="费用信息">
          <span>
            (实际费用){{ orderDetail.order.actualPrice }}元 =
            (商品总价){{ orderDetail.order.goodsPrice }}元 +
            (快递费用){{ orderDetail.order.freightPrice }}元 -
            (优惠减免){{ orderDetail.order.couponPrice }}元 -
            (积分减免){{ orderDetail.order.integralPrice }}元
          </span>
        </el-form-item>
        <el-form-item label="支付信息">
          <span>(支付渠道)微信支付</span>
          <span>(支付时间){{ orderDetail.order.payTime }}</span>
        </el-form-item>
```

```html
            <el-form-item label="快递信息">
                <span>(快递公司){{ orderDetail.order.shipChannel }}</span>
                <span>(快递单号){{ orderDetail.order.shipSn }}</span>
                <span>(发货时间){{ orderDetail.order.shipTime }}</span>
            </el-form-item>
            <el-form-item label="收货信息">
                <span>(确认收货时间){{ orderDetail.order.confirmTime }}</span>
            </el-form-item>
        </el-form>
    </el-dialog>

    <!-- 发货对话框 -->
    <el-dialog :visible.sync="shipDialogVisible" title="发货">
        <el-form ref="shipForm" :model="shipForm" status-icon
                    label-position="left"
                 label-width="100px" style="width: 400px;margin-left:50px;">
            <el-form-item label="快递公司" prop="shipChannel">
                <el-input v-model="shipForm.shipChannel"/>
            </el-form-item>
            <el-form-item label="快递编号" prop="shipSn">
                <el-input v-model="shipForm.shipSn"/>
            </el-form-item>
        </el-form>
        <div slot="footer" class="dialog-footer">
            <el-button @click="shipDialogVisible = false">取消</el-button>
            <el-button type="primary" @click="confirmShip">确定</el-button>
        </div>
    </el-dialog>

    <!-- 退款对话框 -->
    <el-dialog :visible.sync="refundDialogVisible" title="退款">
        <el-form ref="refundForm" :model="refundForm" status-icon
                    label-position="left"
                 label-width="100px" style="width: 400px; margin-left:50px;">
            <el-form-item label="退款金额" prop="refundMoney">
                <el-input v-model="refundForm.refundMoney" :disabled="true"/>
            </el-form-item>
        </el-form>
        <div slot="footer" class="dialog-footer">
            <el-button @click="refundDialogVisible = false">取消</el-button>
            <el-button type="primary" @click="confirmRefund">确定</el-button>
        </div>
    </el-dialog>
  </div>
</template>
```

(2) 在后台订单管理模块中,后端订单管理功能通过视图文件 litemall-admin-api\src\main\

java\org\linlinjava\litemall\admin\web\AdminOrderController.java 实现，主要功能是获取系统数据库中的订单信息，然后将获取的订单信息列表显示在页面中；同时实现对某条订单的处理功能，例如订单详情、订单退款、发货、订单操作结果和回复订单商品等。文件 AdminOrderController.java 的主要代码如下。

```java
public class AdminOrderController {
    private final Log logger = LogFactory.getLog(AdminOrderController.class);

    @Autowired
    private AdminOrderService adminOrderService;
    /**
     * 查询订单
     */
    @RequiresPermissions("admin:order:list")
    @RequiresPermissionsDesc(menu = {"商场管理", "订单管理"}, button = "查询")
    @GetMapping("/list")
    public Object list(Integer userId, String orderSn,
                       @RequestParam(required = false) List<Short> orderStatusArray,
                       @RequestParam(defaultValue = "1") Integer page,
                       @RequestParam(defaultValue = "10") Integer limit,
                       @Sort @RequestParam(defaultValue = "add_time") String sort,
                       @Order @RequestParam(defaultValue = "desc") String order) {
        return adminOrderService.list(userId, orderSn, orderStatusArray,
             page, limit, sort, order);
    }

    /**
     * 订单详情
     */
    @RequiresPermissions("admin:order:read")
    @RequiresPermissionsDesc(menu = {"商场管理", "订单管理"}, button = "详情")
    @GetMapping("/detail")
    public Object detail(@NotNull Integer id) {
        return adminOrderService.detail(id);
    }

    /**
     * 订单退款
     * @param body 订单信息, { orderId: xxx }
     * @return 订单退款操作结果
     */
    @RequiresPermissions("admin:order:refund")
    @RequiresPermissionsDesc(menu = {"商场管理", "订单管理"}, button = "订单退款")
    @PostMapping("/refund")
    public Object refund(@RequestBody String body) {
        return adminOrderService.refund(body);
```

```java
    }

    /**
     * 发货
     * @param body 订单信息, { orderId: xxx, shipSn: xxx, shipChannel: xxx }
     * @return 订单操作结果
     */
    @RequiresPermissions("admin:order:ship")
    @RequiresPermissionsDesc(menu = {"商场管理", "订单管理"}, button = "订单发货")
    @PostMapping("/ship")
    public Object ship(@RequestBody String body) {
        return adminOrderService.ship(body);
    }
    /**
     * 回复订单商品
     * @param body 订单信息, { orderId: xxx }
     * @return 订单操作结果
     */
    @RequiresPermissions("admin:order:reply")
    @RequiresPermissionsDesc(menu = {"商场管理", "订单管理"},
                            button = "订单商品回复")
    @PostMapping("/reply")
    public Object reply(@RequestBody String body) {
        return adminOrderService.reply(body);
    }

}
```

2.4.4 商品管理

后台商品管理模块包含商品列表、商品上架和商品评论 3 个子选项。接下来将详细讲解商品管理模块的实现过程。

(1) 在后台商品管理模块中，前端商品列表页面由文件 litemall\litemall-admin\src\views\goods\list.vue 实现，在此文件顶部显示商品搜索表单和按钮，在下方分页列表显示系统内的所有商品信息和对应的操作按钮。

(2) 在后台商品管理模块中，前端商品上架页面由文件 litemall\litemall-admin\src\views\goods\create.vue 实现，在此页面中显示添加新商品表单。

(3) 在后台商品管理模块中，前端商品评论页面由文件 litemall\litemall-admin\src\views\goods\comment.vue 实现，在此页面中显示用户对系统内商品的所有评价信息。文件 comment.vue 的主要代码如下。

```html
<el-table v-loading="listLoading" :data="list" size="small" element-loading-text=
    "正在查询中…" border fit highlight-current-row>
  <el-table-column align="center" label="用户 ID" prop="userId"/>
  <el-table-column align="center" label="商品 ID" prop="valueId"/>
  <el-table-column align="center" label="打分" prop="star"/>
  <el-table-column align="center" label="评论内容" prop="content"/>
  <el-table-column align="center" label="评论图片" prop="picUrls"/>
    <template slot-scope="scope">
      <img v-for="item in scope.row.picUrls" :key="item" :src="item" width="40">
    </template>
  </el-table-column>
  <el-table-column align="center" label="时间" prop="addTime"/>
  <el-table-column align="center" label="操作" width="200" class-name=
      "small-padding fixed-width">
    <template slot-scope="scope">
      <el-button type="primary" size="mini" @click="handleReply(scope.row)">回复</el-button>
      <el-button type="danger" size="mini" @click="handleDelete(scope.row)">删除</el-button>
    </template>
  </el-table-column>
</el-table>
<pagination v-show="total>0" :total="total" :page.sync="listQuery.page" :
    limit.sync="listQuery.limit" @pagination="getList" />
<!-- 评论回复 -->
<el-dialog :visible.sync="replyFormVisible" title="回复">
  <el-form ref="replyForm" :model="replyForm" status-icon label-position="left"
      label-width="100px" style="width: 400px; margin-left:50px;">
    <el-form-item label="回复内容" prop="content">
      <el-input :autosize="{ minRows: 4, maxRows: 8}" v-model="replyForm.content"
          type="textarea"/>
    </el-form-item>
  </el-form>
  <div slot="footer" class="dialog-footer">
    <el-button @click="replyFormVisible = false">取消</el-button>
    <el-button type="primary" @click="reply">确定</el-button>
  </div>
</el-dialog>
  </div>
</template>
```

(4) 在后台商品管理模块中,当单击商品列表某个商品后面的"编辑"按钮时,会弹出一个修改商品页面,这个页面由文件 litemall\litemall-admin\src\views\goods\edit.vue 实现,主要代码如下。

```html
<template>
  <div class="app-container">
    <el-card class="box-card">
      <h3>商品介绍</h3>
      <el-form ref="goods" :rules="rules" :model="goods" label-width="150px">
        <el-form-item label="商品编号" prop="goodsSn">
          <el-input v-model="goods.goodsSn"/>
        </el-form-item>
        <el-form-item label="商品名称" prop="name">
          <el-input v-model="goods.name"/>
        </el-form-item>
        <el-form-item label="专柜价格" prop="counterPrice">
          <el-input v-model="goods.counterPrice" placeholder="0.00">
            <template slot="append">元</template>
          </el-input>
        </el-form-item>
        <el-form-item label="当前价格" prop="retailPrice">
          <el-input v-model="goods.retailPrice" placeholder="0.00">
            <template slot="append">元</template>
          </el-input>
        </el-form-item>
        <el-form-item label="是否新品" prop="isNew">
          <el-radio-group v-model="goods.isNew">
            <el-radio :label="true">新品</el-radio>
            <el-radio :label="false">非新品</el-radio>
          </el-radio-group>
        </el-form-item>
        <el-form-item label="是否热卖" prop="isHot">
          <el-radio-group v-model="goods.isHot">
            <el-radio :label="false">普通</el-radio>
            <el-radio :label="true">热卖</el-radio>
          </el-radio-group>
        </el-form-item>
        <el-form-item label="是否在售" prop="isOnSale">
          <el-radio-group v-model="goods.isOnSale">
            <el-radio :label="true">在售</el-radio>
            <el-radio :label="false">未售</el-radio>
          </el-radio-group>
        </el-form-item>

        <el-form-item label="商品图片">
          <el-upload
            :headers="headers"
            :action="uploadPath"
            :show-file-list="false"
            :on-success="uploadPicUrl"
            class="avatar-uploader"
```

```
            accept=".jpg,.jpeg,.png,.gif">
            <img v-if="goods.picUrl" :src="goods.picUrl" class="avatar">
            <i v-else class="el-icon-plus avatar-uploader-icon"/>
          </el-upload>
        </el-form-item>

        <el-form-item label="宣传画廊">
          <el-upload
            :action="uploadPath"
            :headers="headers"
            :limit="5"
            :file-list="galleryFileList"
            :on-exceed="uploadOverrun"
            :on-success="handleGalleryUrl"
            :on-remove="handleRemove"
            multiple
            accept=".jpg,.jpeg,.png,.gif"
            list-type="picture-card">
            <i class="el-icon-plus"/>
          </el-upload>
        </el-form-item>
```

(5) 在后台商品管理模块中，后端商品列表功能通过视图文件 litemall-admin-api\src\main\java\org\linlinjava\litemall\admin\web\AdminGoodsController.java 实现，主要功能是获取系统数据库中的商品信息，然后分别实现显示商品列表、添加新商品、修改商品和删除商品等功能。文件 AdminGoodsController.java 的主要代码如下。

```java
@RequestMapping("/admin/goods")
@Validated
public class AdminGoodsController {
    private final Log logger = LogFactory.getLog(AdminGoodsController.class);
    @Autowired
    private AdminGoodsService adminGoodsService;

    /**
     * 查询商品
     */
    @RequiresPermissions("admin:goods:list")
    @RequiresPermissionsDesc(menu = {"商品管理", "商品管理"}, button = "查询")
    @GetMapping("/list")
    public Object list(String goodsSn, String name,
                       @RequestParam(defaultValue = "1") Integer page,
                       @RequestParam(defaultValue = "10") Integer limit,
                       @Sort @RequestParam(defaultValue = "add_time") String sort,
                       @Order @RequestParam(defaultValue = "desc") String order) {
        return adminGoodsService.list(goodsSn, name, page, limit, sort, order);
```

```java
}
@GetMapping("/catAndBrand")
public Object list2() {
    return adminGoodsService.list2();
}
/**
 * 编辑商品
 */
@RequiresPermissions("admin:goods:update")
@RequiresPermissionsDesc(menu = {"商品管理", "商品管理"}, button = "编辑")
@PostMapping("/update")
public Object update(@RequestBody GoodsAllinone goodsAllinone) {
    return adminGoodsService.update(goodsAllinone);
}
/**
 * 删除商品
 */
@RequiresPermissions("admin:goods:delete")
@RequiresPermissionsDesc(menu = {"商品管理", "商品管理"}, button = "删除")
@PostMapping("/delete")
public Object delete(@RequestBody LitemallGoods goods) {
    return adminGoodsService.delete(goods);
}
/**
 * 添加商品
 */
@RequiresPermissions("admin:goods:create")
@RequiresPermissionsDesc(menu = {"商品管理", "商品管理"}, button = "上架")
@PostMapping("/create")
public Object create(@RequestBody GoodsAllinone goodsAllinone) {
    return adminGoodsService.create(goodsAllinone);
}
/**
 * 商品详情
 */
@RequiresPermissions("admin:goods:read")
@RequiresPermissionsDesc(menu = {"商品管理", "商品管理"}, button = "详情")
@GetMapping("/detail")
public Object detail(@NotNull Integer id) {
    return adminGoodsService.detail(id);
}
}
```

2.5 实现小商城系统

本项目的小商城系统模块分为前端和后端两部分，其中前端模块是 litemall-wx，基于微信小程序技术实现；后端模块是 litemall-wx-api，基于 Spring Boot 技术实现。本节将详细讲解小商城系统中主要功能的实现过程。

扫码看视频

2.5.1 系统主页

小商城系统模块的主页前端由文件 litemall\litemall-wx\pages\index\index.wxml 实现，功能是展示微信商城的主页信息，主要代码如下。

```
<view class="container">
 <swiper class="goodsimgs" indicator-dots="true" autoplay="true" interval="3000" duration="1000">
   <swiper-item wx:for="{{goods.gallery}}" wx:key="*this">
     <image src="{{item}}" background-size="cover"></image>
   </swiper-item>
 </swiper>
<!-- 分享 -->
<view class='goods_name'>
   <view class='goods_name_left'>{{goods.name}}</view>
   <view class="goods_name_right" bindtap="shareFriendOrCircle">分享</view>
</view>
<view class="share-pop-box" hidden="{{!openShare}}">
   <view class="share-pop">
      <view class="close" bindtap="closeShare">
        <image class="icon" src="/static/images/icon_close.png"></image>
      </view>
      <view class='share-info'>
      <button class="sharebtn" open-type="share" wx:if="{{!isGroupon}}">
         <image class='sharebtn_image' src='/static/images/wechat.png'></image>
         <view class='sharebtn_text'>分享给好友</view>
      </button>
      <button class="savesharebtn" open-type="openSetting" bindopensetting=
         "handleSetting" wx:if="{{(!isGroupon) && (!canWrite)}}" >
         <image class='sharebtn_image' src='/static/images/friend.png'></image>
         <view class='sharebtn_text'>发朋友圈</view>
      </button>
      <button class="savesharebtn" bindtap="saveShare" wx:if="{{!isGroupon && canWrite}}">
         <image class='sharebtn_image' src='/static/images/friend.png'></image>
         <view class='sharebtn_text'>发朋友圈</view>
      </button>
```

```
            </view>
          </view>
      </view>
      <view class="goods-info">
        <view class="c">
        <text class="desc">{{goods.goodsBrief}}</text>
        <view class="price">
            <view class="counterPrice">原价: ¥{{goods.counterPrice}}</view>
            <view class="retailPrice">现价: ¥{{checkedSpecPrice}}</view>
        </view>
        <view class="brand" wx:if="{{brand.name}}">
          <navigator url="../brandDetail/brandDetail?id={{brand.id}}">
            <text>{{brand.name}}</text>
          </navigator>
         </view>
        </view>
      </view>
      <view class="section-nav section-attr" bindtap="switchAttrPop">
          <view class="t">{{checkedSpecText}}</view>
          <image class="i" src="/static/images/address_right.png" background-size="cover"></image>
       </view>
       <view class="comments" wx:if="{{comment.count > 0}}">
          <view class="h">
            <navigator url="/pages/comment/comment?valueId={{goods.id}}&type=0">
              <text class="t">评价({{comment.count > 999 ? '999+' : comment.count}})</text>
              <text class="i">查看全部</text>
            </navigator>
          </view>
          <view class="b">
              <view class="item" wx:for="{{comment.data}}" wx:key="id">
                  <view class="info">
                      <view class="user">
                          <image src="{{item.avatar}}"></image>
                          <text>{{item.nickname}}</text>
                      </view>
                      <view class="time">{{item.addTime}}</view>
                  </view>
                  <view class="content">
                     {{item.content}}
                  </view>
                  <view class="imgs" wx:if="{{item.picList.length > 0}}">
                     <image class="img" wx:for="{{item.picList}}" wx:key="*this"
                         wx:for-item="item" src="{{item}} "></image>
                  </view>
              </view>
           </view>
        </view>
```

2.5.2 会员注册登录

(1) 小商城系统模块的会员注册前端由文件 litemall\litemall-wx\pages\auth\register.wxml 实现，功能是实现新用户的注册，并在注册成功后将注册信息添加到系统中。文件 register.wxml 的主要代码如下。

```
<view class="container">
  <view class="form-box">
    <view class="form-item">
      <input class="username" value="{{username}}" bindinput="bindUsernameInput" placeholder="用户名" auto-focus/>
      <image wx:if="{{ username.length > 0 }}" id="clear-username" class="clear" src="/static/images/clear_input.png" catchtap="clearInput"></image>
    </view>

    <view class="form-item">
      <input class="password" value="{{password}}" password bindinput="bindPasswordInput" placeholder="密码" />
      <image class="clear" id="clear-password" wx:if="{{ password.length > 0 }}" src="/static/images/clear_input.png" catchtap="clearInput"></image>
    </view>
    <view class="form-item">
      <input class="password" value="{{confirmPassword}}" password bindinput="bindConfirmPasswordInput" placeholder="确认密码" />
      <image class="clear" id="clear-confirm-password" wx:if="{{ confirmPassword.length > 0 }}" src="/static/images/clear_input.png" catchtap="clearInput"></image>
    </view>

    <view class="form-item">
      <input class="mobile" value="{{mobile}}" bindinput="bindMobileInput" placeholder="手机号" />
      <image wx:if="{{ mobile.length > 0 }}" id="clear-mobile" class="clear" src="/static/images/clear_input.png" catchtap="clearInput"></image>
    </view>
    <view class="form-item-code">
      <view class="form-item code-item">
        <input class="code" value="{{code}}" bindinput="bindCodeInput" placeholder="验证码" />
        <image class="clear" id="clear-code" wx:if="{{ code.length > 0 }}" src="/static/images/clear_input.png" catchtap="clearInput"></image>
      </view>
      <view class="code-btn" bindtap="sendCode">获取验证码</view>
    </view>
```

```
    <button type="primary" class="register-btn" bindtap="startRegister">注册
</button>
  </view>
</view>
```

(2) 小商城系统模块的会员登录前端由文件 litemall\litemall-wx\pages\auth\login.wxml 实现，功能是实现会员用户的登录验证。文件 login.wxml 的主要代码如下。

```
<view class="container">
  <view class="login-box">
    <button type="primary" open-type="getUserInfo" class="wx-login-btn" bindgetuserinfo="wxLogin">微信直接登录</button>
    <button type="primary" class="account-login-btn" bindtap="accountLogin">账号登录</button>
  </view>
</view>
```

(3) 小商城系统模块的会员注册和登录验证后端由文件 litemall\litemall-wx-api\src\main\java\org\linlinjava\litemall\wx\web\WxAuthController.java 实现，具体实现流程如下。

① 实现注册功能：获取注册信息，并将合法的注册信息保存到系统。主要代码如下。

```
@PostMapping("register")
public Object register(@RequestBody String body, HttpServletRequest request) {
    String username = JacksonUtil.parseString(body, "username");
    String password = JacksonUtil.parseString(body, "password");
    String mobile = JacksonUtil.parseString(body, "mobile");
    String code = JacksonUtil.parseString(body, "code");
    String wxCode = JacksonUtil.parseString(body, "wxCode");

    if (StringUtils.isEmpty(username) || StringUtils.isEmpty(password)
            || StringUtils.isEmpty(mobile) || StringUtils.isEmpty(wxCode)
            || StringUtils.isEmpty(code)) {
        return ResponseUtil.badArgument();
    }
    List<LitemallUser> userList = userService.queryByUsername(username);
    if (userList.size() > 0) {
        return ResponseUtil.fail(AUTH_NAME_REGISTERED, "用户名已注册");
    }
    userList = userService.queryByMobile(mobile);
    if (userList.size() > 0) {
        return ResponseUtil.fail(AUTH_MOBILE_REGISTERED, "手机号已注册");
    }
    if (!RegexUtil.isMobileExact(mobile)) {
        return ResponseUtil.fail(AUTH_INVALID_MOBILE, "手机号格式不正确");
    }
    //判断验证码是否正确
    String cacheCode = CaptchaCodeManager.getCachedCaptcha(mobile);
```

```java
        if (cacheCode == null || cacheCode.isEmpty() || !cacheCode.equals(code)) {
            return ResponseUtil.fail(AUTH_CAPTCHA_UNMATCH, "验证码错误");
        }
        String openId = null;
        try {
            WxMaJscode2SessionResult result =
                        this.wxService.getUserService().getSessionInfo(wxCode);
            openId = result.getOpenid();
        } catch (Exception e) {
            e.printStackTrace();
            return ResponseUtil.fail(AUTH_OPENID_UNACCESS, "openid 获取失败");
        }
        userList = userService.queryByOpenid(openId);
        if (userList.size() > 1) {
            return ResponseUtil.serious();
        }
        if (userList.size() == 1) {
            LitemallUser checkUser = userList.get(0);
            String checkUsername = checkUser.getUsername();
            String checkPassword = checkUser.getPassword();
            if (!checkUsername.equals(openId) || !checkPassword.equals(openId)) {
                return ResponseUtil.fail(AUTH_OPENID_BINDED, "openid已绑定账号");
            }
        }
        LitemallUser user = null;
        BCryptPasswordEncoder encoder = new BCryptPasswordEncoder();
        String encodedPassword = encoder.encode(password);
        user = new LitemallUser();
        user.setUsername(username);
        user.setPassword(encodedPassword);
        user.setMobile(mobile);
        user.setWeixinOpenid(openId);
        user.setAvatar("https://yanxuan.nosdn.127.net/
80841d741d7fa3073e0ae27bf487339f.jpg?imageView&quality=90&thumbnail=64x64");
        user.setNickname(username);
        user.setGender((byte) 0);
        user.setUserLevel((byte) 0);
        user.setStatus((byte) 0);
        user.setLastLoginTime(LocalDateTime.now());
        user.setLastLoginIp(IpUtil.client(request));
        userService.add(user);

        // 给新用户发送注册优惠券
        couponAssignService.assignForRegister(user.getId());
        // userInfo
        UserInfo userInfo = new UserInfo();
        userInfo.setNickName(username);
```

```java
        userInfo.setAvatarUrl(user.getAvatar());
        // token
        UserToken userToken = UserTokenManager.generateToken(user.getId());

        Map<Object, Object> result = new HashMap<Object, Object>();
        result.put("token", userToken.getToken());
        result.put("tokenExpire", userToken.getExpireTime().toString());
        result.put("userInfo", userInfo);
        return ResponseUtil.ok(result);
    }
```

② 实现登录验证功能:获取登录表单中的信息,然后验证登录信息的合法性。本系统支持微信登录和手机发送验证码登录,主要代码如下。

```java
/**
 * 鉴权服务
 */
@RestController
@RequestMapping("/wx/auth")
@Validated
public class WxAuthController {
    private final Log logger = LogFactory.getLog(WxAuthController.class);

    @Autowired
    private LitemallUserService userService;

    @Autowired
    private WxMaService wxService;

    @Autowired
    private NotifyService notifyService;

    @Autowired
    private CouponAssignService couponAssignService;

    /**
     * 账号登录
     *
     * @param body    请求内容,{ username: xxx, password: xxx }
     * @param request 请求对象
     * @return 登录结果
     */
    @PostMapping("login")
    public Object login(@RequestBody String body, HttpServletRequest request) {
        String username = JacksonUtil.parseString(body, "username");
        String password = JacksonUtil.parseString(body, "password");
        if (username == null || password == null) {
```

```java
        return ResponseUtil.badArgument();
    }

    List<LitemallUser> userList = userService.queryByUsername(username);
    LitemallUser user = null;
    if (userList.size() > 1) {
        return ResponseUtil.serious();
    } else if (userList.size() == 0) {
        return ResponseUtil.badArgumentValue();
    } else {
        user = userList.get(0);
    }

    BCryptPasswordEncoder encoder = new BCryptPasswordEncoder();
    if (!encoder.matches(password, user.getPassword())) {
        return ResponseUtil.fail(AUTH_INVALID_ACCOUNT, "账号密码不对");
    }

    // userInfo
    UserInfo userInfo = new UserInfo();
    userInfo.setNickName(username);
    userInfo.setAvatarUrl(user.getAvatar());

    // token
    UserToken userToken = UserTokenManager.generateToken(user.getId());

    Map<Object, Object> result = new HashMap<Object, Object>();
    result.put("token", userToken.getToken());
    result.put("tokenExpire", userToken.getExpireTime().toString());
    result.put("userInfo", userInfo);
    return ResponseUtil.ok(result);
}

/**
 * 微信登录
 *
 * @param wxLoginInfo 请求内容,{ code: xxx, userInfo: xxx }
 * @param request     请求对象
 * @return 登录结果
 */
@PostMapping("login_by_weixin")
public Object loginByWeixin(@RequestBody WxLoginInfo wxLoginInfo,
                            HttpServletRequest request) {
    String code = wxLoginInfo.getCode();
    UserInfo userInfo = wxLoginInfo.getUserInfo();
    if (code == null || userInfo == null) {
        return ResponseUtil.badArgument();
```

```java
        }
        String sessionKey = null;
        String openId = null;
        try {
            WxMaJscode2SessionResult result =
                        this.wxService.getUserService().getSessionInfo(code);
            sessionKey = result.getSessionKey();
            openId = result.getOpenid();
        } catch (Exception e) {
            e.printStackTrace();
        }

        if (sessionKey == null || openId == null) {
            return ResponseUtil.fail();
        }

        LitemallUser user = userService.queryByOid(openId);
        if (user == null) {
            user = new LitemallUser();
            user.setUsername(openId);
            user.setPassword(openId);
            user.setWeixinOpenid(openId);
            user.setAvatar(userInfo.getAvatarUrl());
            user.setNickname(userInfo.getNickName());
            user.setGender(userInfo.getGender());
            user.setUserLevel((byte) 0);
            user.setStatus((byte) 0);
            user.setLastLoginTime(LocalDateTime.now());
            user.setLastLoginIp(IpUtil.client(request));

            userService.add(user);

            // 新用户发送注册优惠券
            couponAssignService.assignForRegister(user.getId());
        } else {
            user.setLastLoginTime(LocalDateTime.now());
            user.setLastLoginIp(IpUtil.client(request));
            if (userService.updateById(user) == 0) {
                return ResponseUtil.updatedDataFailed();
            }
        }
        UserToken userToken = UserTokenManager.generateToken(user.getId());
        userToken.setSessionKey(sessionKey);

        Map<Object, Object> result = new HashMap<Object, Object>();
        result.put("token", userToken.getToken());
```

```java
        result.put("tokenExpire", userToken.getExpireTime().toString());
        result.put("userInfo", userInfo);
        return ResponseUtil.ok(result);
}
/**
 * 请求验证码
 * @param body 手机号码{mobile}
 */
@PostMapping("regCaptcha")
public Object registerCaptcha(@RequestBody String body) {
    String phoneNumber = JacksonUtil.parseString(body, "mobile");
    if (StringUtils.isEmpty(phoneNumber)) {
        return ResponseUtil.badArgument();
    }
    if (!RegexUtil.isMobileExact(phoneNumber)) {
        return ResponseUtil.badArgumentValue();
    }

    if (!notifyService.isSmsEnable()) {
        return ResponseUtil.fail(AUTH_CAPTCHA_UNSUPPORT, "小程序后台验证码服务不支持");
    }
    String code = CharUtil.getRandomNum(6);
    notifyService.notifySmsTemplate(phoneNumber, NotifyType.CAPTCHA, new
                                    String[]{code});

    boolean successful = CaptchaCodeManager.addToCache(phoneNumber, code);
    if (!successful) {
        return ResponseUtil.fail(AUTH_CAPTCHA_FREQUENCY, "验证码未超时1分钟,不能发送");
    }

    return ResponseUtil.ok();
}
```

③ 实现密码重置功能:在重置密码时,需要用到手机验证码,主要代码如下。

```java
@PostMapping("reset")
public Object reset(@RequestBody String body, HttpServletRequest request) {
    String password = JacksonUtil.parseString(body, "password");
    String mobile = JacksonUtil.parseString(body, "mobile");
    String code = JacksonUtil.parseString(body, "code");

    if (mobile == null || code == null || password == null) {
        return ResponseUtil.badArgument();
    }
    //判断验证码是否正确
    String cacheCode = CaptchaCodeManager.getCachedCaptcha(mobile);
    if (cacheCode == null || cacheCode.isEmpty() || !cacheCode.equals(code))
        return ResponseUtil.fail(AUTH_CAPTCHA_UNMATCH, "验证码错误");
```

```
        List<LitemallUser> userList = userService.queryByMobile(mobile);
        LitemallUser user = null;
        if (userList.size() > 1) {
            return ResponseUtil.serious();
        } else if (userList.size() == 0) {
            return ResponseUtil.fail(AUTH_MOBILE_UNREGISTERED, "手机号未注册");
        } else {
            user = userList.get(0);
        }
        BCryptPasswordEncoder encoder = new BCryptPasswordEncoder();
        String encodedPassword = encoder.encode(password);
        user.setPassword(encodedPassword);

        if (userService.updateById(user) == 0) {
            return ResponseUtil.updatedDataFailed();
        }
        return ResponseUtil.ok();
    }

    @PostMapping("bindPhone")
    public Object bindPhone(@LoginUser Integer userId, @RequestBody String body) {
        String sessionKey = UserTokenManager.getSessionKey(userId);
        String encryptedData = JacksonUtil.parseString(body, "encryptedData");
        String iv = JacksonUtil.parseString(body, "iv");
        WxMaPhoneNumberInfo phoneNumberInfo = this.wxService.getUserService().
                    getPhoneNoInfo(sessionKey, encryptedData, iv);
        String phone = phoneNumberInfo.getPhoneNumber();
        LitemallUser user = userService.findById(userId);
        user.setMobile(phone);
        if (userService.updateById(user) == 0) {
            return ResponseUtil.updatedDataFailed();
        }
        return ResponseUtil.ok();
    }
```

2.5.3 商品分类

(1) 小商城系统模块的商品分类前端由文件 litemall\litemall-wx\pages\catalog\catalog.wxml 实现，功能是为了方便用户快速找到自己需要的商品，将系统内的商品进行分类。文件 catalog.wxml 的主要代码如下：

```
<view class="catalog">
  <scroll-view class="nav" scroll-y="true">
```

```
      <view class="item {{ currentCategory.id == item.id ? 'active' : ''}}"
wx:for="{{categoryList}}" wx:key="id" data-id="{{item.id}}"
data-index="{{index}}" bindtap="switchCate">{{item.name}}</view>
    </scroll-view>
    <scroll-view class="cate" scroll-y="true">
      <navigator url="url" class="banner">
        <image class="image" src="{{currentCategory.picUrl}}"></image>
        <view class="txt">{{currentCategory.frontName}}</view>
      </navigator>
      <view class="hd">
        <text class="line"></text>
        <text class="txt">{{currentCategory.name}}分类</text>
        <text class="line"></text>
      </view>
      <view class="bd">
        <navigator url="/pages/category/category?id={{item.id}}" class="item
{{(index+1) % 3 == 0 ? 'last' : ''}}" wx:key="id" wx:for="{{currentSubCategoryList}}">
          <image class="icon" src="{{item.picUrl}}"></image>
          <text class="txt">{{item.name}}</text>
        </navigator>
      </view>
    </scroll-view>
  </view>
</view>
```

(2) 小商城系统模块的商品分类后端由文件 litemall\litemall-wx-api\src\main\java\org\linlinjava\litemall\wx\web\WxCatalogController.java 实现，主要代码如下。

```
public class WxCatalogController {
    private final Log logger = LogFactory.getLog(WxCatalogController.class);

    @Autowired
    private LitemallCategoryService categoryService;

    /**
     * 分类详情
     *
     * @param id    分类类目 ID。
     *              如果分类类目 ID 是空，则选择第一个分类类目。
     *              需要注意，这里的分类类目是一级类目
     * @return 分类详情
     */
    @GetMapping("index")
    public Object index(Integer id) {

        // 所有一级分类目录
        List<LitemallCategory> l1CatList = categoryService.queryL1();
```

```java
        // 当前一级分类目录
        LitemallCategory currentCategory = null;
        if (id != null) {
            currentCategory = categoryService.findById(id);
        } else {
            currentCategory = l1CatList.get(0);
        }

        // 当前一级分类目录对应的二级分类目录
        List<LitemallCategory> currentSubCategory = null;
        if (null != currentCategory) {
            currentSubCategory = categoryService.queryByPid(currentCategory.getId());
        }

        Map<String, Object> data = new HashMap<String, Object>();
        data.put("categoryList", l1CatList);
        data.put("currentCategory", currentCategory);
        data.put("currentSubCategory", currentSubCategory);
        return ResponseUtil.ok(data);
    }

    /**
     * 所有分类数据
     *
     * @return 所有分类数据
     */
    @GetMapping("all")
    public Object queryAll() {
        //优先从缓存中读取
        if (HomeCacheManager.hasData(HomeCacheManager.CATALOG)) {
            return ResponseUtil.ok(HomeCacheManager.getCacheData
                            (HomeCacheManager.CATALOG));
        }
        //所有一级分类目录
        List<LitemallCategory> l1CatList = categoryService.queryL1();

        //所有子分类列表
        Map<Integer, List<LitemallCategory>> allList = new HashMap<>();
        List<LitemallCategory> sub;
        for (LitemallCategory category : l1CatList) {
            sub = categoryService.queryByPid(category.getId());
            allList.put(category.getId(), sub);
        }
        //当前一级分类目录
        LitemallCategory currentCategory = l1CatList.get(0);
        //当前一级分类目录对应的二级分类目录
```

```
            List<LitemallCategory> currentSubCategory = null;
            if (null != currentCategory) {
                currentSubCategory = categoryService.queryByPid(currentCategory.getId());
            }
            Map<String, Object> data = new HashMap<String, Object>();
            data.put("categoryList", l1CatList);
            data.put("allList", allList);
            data.put("currentCategory", currentCategory);
            data.put("currentSubCategory", currentSubCategory);

            //缓存数据
            HomeCacheManager.loadData(HomeCacheManager.CATALOG, data);
            return ResponseUtil.ok(data);
    }

    /**
     * 当前分类栏目
     * @param id 分类类目 ID
     * @return 当前分类栏目
     */
    @GetMapping("current")
    public Object current(@NotNull Integer id) {
        //当前分类
        LitemallCategory currentCategory = categoryService.findById(id);
        List<LitemallCategory> currentSubCategory = categoryService.queryByPid(currentCategory.getId());

        Map<String, Object> data = new HashMap<String, Object>();
        data.put("currentCategory", currentCategory);
        data.put("currentSubCategory", currentSubCategory);
        return ResponseUtil.ok(data);
    }
}
```

2.5.4 商品搜索

(1) 小商城系统模块的商品搜索前端由文件 litemall\litemall-wx\pages\search\search.wxml 实现，功能是为了方便用户快速找到自己需要的商品，可以在搜索表单中输入关键字进行搜索。

(2) 小商城系统模块的商品搜索后端由文件 litemall\litemall-wx-api\src\main\java\org\linlinjava\litemall\wx\web\WxSearchController.java 实现，主要代码如下。

```
public class WxSearchController {
    private final Log logger = LogFactory.getLog(WxSearchController.class);
```

```java
@Autowired
private LitemallKeywordService keywordsService;
@Autowired
private LitemallSearchHistoryService searchHistoryService;

/**
 * 搜索页面信息
 * 如果用户已登录，则给出用户历史搜索记录；
 * 如果没有登录，则给出空历史搜索记录。
 * @param userId 用户ID，可选
 * @return 搜索页面信息
 */
@GetMapping("index")
public Object index(@LoginUser Integer userId) {
    //取出输入框默认的关键字
    LitemallKeyword defaultKeyword = keywordsService.queryDefault();
    //取出热门关键字
    List<LitemallKeyword> hotKeywordList = keywordsService.queryHots();

    List<LitemallSearchHistory> historyList = null;
    if (userId != null) {
        //取出用户历史关键字
        historyList = searchHistoryService.queryByUid(userId);
    } else {
        historyList = new ArrayList<>(0);
    }

    Map<String, Object> data = new HashMap<String, Object>();
    data.put("defaultKeyword", defaultKeyword);
    data.put("historyKeywordList", historyList);
    data.put("hotKeywordList", hotKeywordList);
    return ResponseUtil.ok(data);
}

/**
 * 关键字提醒
 * 当用户输入关键字一部分时，可以推荐系统中匹配的关键字。
 * @param keyword 关键字
 * @return 匹配的关键字
 */
@GetMapping("helper")
public Object helper(@NotEmpty String keyword,
                     @RequestParam(defaultValue = "1") Integer page,
                     @RequestParam(defaultValue = "10") Integer size) {
    List<LitemallKeyword> keywordsList =
            keywordsService.queryByKeyword(keyword, page, size);
```

```
        String[] keys = new String[keywordsList.size()];
        int index = 0;
        for (LitemallKeyword key : keywordsList) {
            keys[index++] = key.getKeyword();
        }
        return ResponseUtil.ok(keys);
    }

    /**
     * 清除用户搜索历史
     * @param userId 用户 ID
     * @return 清理是否成功
     */
    @PostMapping("clearhistory")
    public Object clearhistory(@LoginUser Integer userId) {
        if (userId == null) {
            return ResponseUtil.unlogin();
        }

        searchHistoryService.deleteByUid(userId);
        return ResponseUtil.ok();
    }
}
```

2.5.5 商品团购

(1) 文件 litemall\litemall-wx\pages\groupon\grouponList\grouponList.wxml 用于实现商品团购列表前端，功能是列表显示可参加的团购信息。

(2) 文件 litemall\litemall-wx\pages\groupon\grouponList\grouponDetail.wxm 用于实现商品团购详情列表前端，功能是显示某团购的详细信息。

(3) 文件 litemall\litemall-wx\pages\groupon\myGroupon\myGroupon.wxml 用于实现我的团购信息前端，功能是显示当前用户的团购信息。

(4) 小商城系统模块的团购功能后端由文件 litemall\litemall-wx-api\src\main\java\org\linlinjava\litemall\wx\web\WxGrouponController.java 实现，具体流程如下。

① 用分页列表的形式显示团购信息，对应代码如下。

```
    /**
     * 团购规则列表
     *
     * @param page 分页页数
     * @param size 分页大小
     * @return 团购规则列表
     */
```

```java
@GetMapping("list")
public Object list(@RequestParam(defaultValue = "1") Integer page,
            @RequestParam(defaultValue = "10") Integer size,
            @Sort @RequestParam(defaultValue = "add_time") String sort,
            @Order @RequestParam(defaultValue = "desc") String order) {
    List<Map<String, Object>> topicList = grouponRulesService.queryList(page,
                                            size, sort, order);
    long total = PageInfo.of(topicList).getTotal();
    Map<String, Object> data = new HashMap<String, Object>();
    data.put("data", topicList);
    data.put("count", total);
    return ResponseUtil.ok(data);
}
```

② 编写函数 detail()，展示某团购的详细信息，对应代码如下。

```java
/**
 * 团购活动详情
 *
 * @param userId     用户 ID
 * @param grouponId  团购活动 ID
 * @return 团购活动详情
 */
@GetMapping("detail")
public Object detail(@LoginUser Integer userId, @NotNull Integer grouponId) {
    if (userId == null) {
        return ResponseUtil.unlogin();
    }

    LitemallGroupon groupon = grouponService.queryById(grouponId);
    if (groupon == null) {
        return ResponseUtil.badArgumentValue();
    }

    LitemallGrouponRules rules = rulesService.queryById(groupon.getRulesId());
    if (rules == null) {
        return ResponseUtil.badArgumentValue();
    }

    //订单信息
    LitemallOrder order = orderService.findById(groupon.getOrderId());
    if (null == order) {
        return ResponseUtil.fail(ORDER_UNKNOWN, "订单不存在");
    }
    if (!order.getUserId().equals(userId)) {
        return ResponseUtil.fail(ORDER_INVALID, "不是当前用户的订单");
    }
    Map<String, Object> orderVo = new HashMap<String, Object>();
```

```
orderVo.put("id", order.getId());
orderVo.put("orderSn", order.getOrderSn());
orderVo.put("addTime", order.getAddTime());
orderVo.put("consignee", order.getConsignee());
orderVo.put("mobile", order.getMobile());
orderVo.put("address", order.getAddress());
orderVo.put("goodsPrice", order.getGoodsPrice());
orderVo.put("freightPrice", order.getFreightPrice());
orderVo.put("actualPrice", order.getActualPrice());
orderVo.put("orderStatusText", OrderUtil.orderStatusText(order));
orderVo.put("handleOption", OrderUtil.build(order));
orderVo.put("expCode", order.getShipChannel());
orderVo.put("expNo", order.getShipSn());

List<LitemallOrderGoods> orderGoodsList =
                    orderGoodsService.queryByOid(order.getId());
List<Map<String, Object>> orderGoodsVoList = new
                    ArrayList<>(orderGoodsList.size());
for (LitemallOrderGoods orderGoods : orderGoodsList) {
    Map<String, Object> orderGoodsVo = new HashMap<>();
    orderGoodsVo.put("id", orderGoods.getId());
    orderGoodsVo.put("orderId", orderGoods.getOrderId());
    orderGoodsVo.put("goodsId", orderGoods.getGoodsId());
    orderGoodsVo.put("goodsName", orderGoods.getGoodsName());
    orderGoodsVo.put("number", orderGoods.getNumber());
    orderGoodsVo.put("retailPrice", orderGoods.getPrice());
    orderGoodsVo.put("picUrl", orderGoods.getPicUrl());
    orderGoodsVo.put("goodsSpecificationValues",
                orderGoods.getSpecifications());
    orderGoodsVoList.add(orderGoodsVo);
}

Map<String, Object> result = new HashMap<>();
result.put("orderInfo", orderVo);
result.put("orderGoods", orderGoodsVoList);

//订单状态为已发货且物流信息不为空
//"YTO", "800669400640887922"
if (order.getOrderStatus().equals(OrderUtil.STATUS_SHIP)) {
    ExpressInfo ei = expressService.getExpressInfo(order.getShipChannel(),
                order.getShipSn());
    result.put("expressInfo", ei);
}

UserVo creator = userService.findUserVoById(groupon.getCreatorUserId());
List<UserVo> joiners = new ArrayList<>();
joiners.add(creator);
```

```java
    int linkGrouponId;
    //这是一个团购发起记录
    if (groupon.getGrouponId() == 0) {
        linkGrouponId = groupon.getId();
    } else {
        linkGrouponId = groupon.getGrouponId();
    }

    List<LitemallGroupon> groupons = grouponService.queryJoinRecord(linkGrouponId);

    UserVo joiner;
    for (LitemallGroupon grouponItem : groupons) {
        joiner = userService.findUserVoById(grouponItem.getUserId());
        joiners.add(joiner);
    }

    result.put("linkGrouponId", linkGrouponId);
    result.put("creator", creator);
    result.put("joiners", joiners);
    result.put("groupon", groupon);
    result.put("rules", rules);
    return ResponseUtil.ok(result);
}
```

③ 编写函数 join(),实现参加某次团购的功能,对应代码如下。

```java
/**
 * 参加团购
 * @param grouponId 团购活动 ID
 * @return 操作结果
 */
@GetMapping("join")
public Object join(@NotNull Integer grouponId) {
    LitemallGroupon groupon = grouponService.queryById(grouponId);
    if (groupon == null) {
        return ResponseUtil.badArgumentValue();
    }
    LitemallGrouponRules rules = rulesService.queryById(groupon.getRulesId());
    if (rules == null) {
        return ResponseUtil.badArgumentValue();
    }
    LitemallGoods goods = goodsService.findById(rules.getGoodsId());
    if (goods == null) {
        return ResponseUtil.badArgumentValue();
    }
    Map<String, Object> result = new HashMap<>();
    result.put("groupon", groupon);
    result.put("goods", goods);
```

```
        return ResponseUtil.ok(result);
    }
```

④ 编写函数 my()，展示当前用户参加团购的信息，对应代码如下。

```
/**
 * 用户开团或入团情况
 *
 * @param userId 用户 ID
 * @param showType 显示类型，如果是 0，就是当前用户发起的团购；否则，就是当前用户参加的团购
 * @return 用户开团或入团情况
 */
@GetMapping("my")
public Object my(@LoginUser Integer userId, @RequestParam(defaultValue = "0") Integer showType) {
    if (userId == null) {
        return ResponseUtil.unlogin();
    }
    List<LitemallGroupon> myGroupons;
    if (showType == 0) {
        myGroupons = grouponService.queryMyGroupon(userId);
    } else {
        myGroupons = grouponService.queryMyJoinGroupon(userId);
    }

    List<Map<String, Object>> grouponVoList = new ArrayList<>(myGroupons.size());

    LitemallOrder order;
    LitemallGrouponRules rules;
    LitemallUser creator;
    for (LitemallGroupon groupon : myGroupons) {
        order = orderService.findById(groupon.getOrderId());
        rules = rulesService.queryById(groupon.getRulesId());
        creator = userService.findById(groupon.getCreatorUserId());

        Map<String, Object> grouponVo = new HashMap<>();
        //填充团购信息
        grouponVo.put("id", groupon.getId());
        grouponVo.put("groupon", groupon);
        grouponVo.put("rules", rules);
        grouponVo.put("creator", creator.getNickname());

        int linkGrouponId;
        //这是一个团购发起记录
        if (groupon.getGrouponId() == 0) {
            linkGrouponId = groupon.getId();
            grouponVo.put("isCreator", creator.getId() == userId);
        } else {
```

```java
            linkGrouponId = groupon.getGrouponId();
            grouponVo.put("isCreator", false);
        }
        int joinerCount = grouponService.countGroupon(linkGrouponId);
        grouponVo.put("joinerCount", joinerCount + 1);

        //填充订单信息
        grouponVo.put("orderId", order.getId());
        grouponVo.put("orderSn", order.getOrderSn());
        grouponVo.put("actualPrice", order.getActualPrice());
        grouponVo.put("orderStatusText", OrderUtil.orderStatusText(order));
        grouponVo.put("handleOption", OrderUtil.build(order));

        List<LitemallOrderGoods> orderGoodsList =
            orderGoodsService.queryByOid(order.getId());
        List<Map<String, Object>> orderGoodsVoList = new
            ArrayList<>(orderGoodsList.size());
        for (LitemallOrderGoods orderGoods : orderGoodsList) {
            Map<String, Object> orderGoodsVo = new HashMap<>();
            orderGoodsVo.put("id", orderGoods.getId());
            orderGoodsVo.put("goodsName", orderGoods.getGoodsName());
            orderGoodsVo.put("number", orderGoods.getNumber());
            orderGoodsVo.put("picUrl", orderGoods.getPicUrl());
            orderGoodsVoList.add(orderGoodsVo);
        }
        grouponVo.put("goodsList", orderGoodsVoList);
        grouponVoList.add(grouponVo);
    }

    Map<String, Object> result = new HashMap<>();
    result.put("count", grouponVoList.size());
    result.put("data", grouponVoList);

    return ResponseUtil.ok(result);
}
```

⑤ 编写函数 query()，展示某商品所对应的团购，对应代码如下。

```java
/**
 * 商品所对应的团购规则
 * @param goodsId 商品ID
 * @return 团购规则详情
 */
@GetMapping("query")
public Object query(@NotNull Integer goodsId) {
    LitemallGoods goods = goodsService.findById(goodsId);
    if (goods == null) {
        return ResponseUtil.fail(GOODS_UNKNOWN, "未找到对应的商品");
```

```
        List<LitemallGrouponRules> rules = rulesService.queryByGoodsId(goodsId);
        return ResponseUtil.ok(rules);
    }
```

2.5.6 购物车

(1) 小商城系统模块的购物车前端由文件 litemall\litemall-wx\pages\cart\cart.wxml 实现，功能是展示购物车中的商品信息，并在购物车中分别实现添加商品和编辑商品功能。

(2) 小商城系统模块的购物车后端由文件 litemall\litemall-wx-api\src\main\java\org\linlinjava\litemall\wx\web\WxCartController.java 实现，主要代码如下。

```java
/**
 * 用户购物车信息
 * @param userId 用户 ID
 * @return 用户购物车信息
 */
@GetMapping("index")
public Object index(@LoginUser Integer userId) {
    if (userId == null) {
        return ResponseUtil.unlogin();
    }
    List<LitemallCart> cartList = cartService.queryByUid(userId);
    Integer goodsCount = 0;
    BigDecimal goodsAmount = new BigDecimal(0.00);
    Integer checkedGoodsCount = 0;
    BigDecimal checkedGoodsAmount = new BigDecimal(0.00);
    for (LitemallCart cart : cartList) {
        goodsCount += cart.getNumber();
        goodsAmount = goodsAmount.add(cart.getPrice().multiply
            (new BigDecimal(cart.getNumber())));
        if (cart.getChecked()) {
            checkedGoodsCount += cart.getNumber();
            checkedGoodsAmount = checkedGoodsAmount.add(cart.getPrice().
                multiply(new BigDecimal(cart.getNumber())));
        }
    }
    Map<String, Object> cartTotal = new HashMap<>();
    cartTotal.put("goodsCount", goodsCount);
    cartTotal.put("goodsAmount", goodsAmount);
    cartTotal.put("checkedGoodsCount", checkedGoodsCount);
    cartTotal.put("checkedGoodsAmount", checkedGoodsAmount);

    Map<String, Object> result = new HashMap<>();
    result.put("cartList", cartList);
```

```java
            result.put("cartTotal", cartTotal);

            return ResponseUtil.ok(result);
}

/**
 * 加入商品到购物车
 * <p>
 * 如果已经存在购物车货品,则增加数量;
 * 否则添加新的购物车货品项。
 *
 * @param userId 用户ID
 * @param cart    购物车商品信息, { goodsId: xxx, productId: xxx, number: xxx }
 * @return 加入购物车操作结果
 */
@PostMapping("add")
public Object add(@LoginUser Integer userId, @RequestBody LitemallCart cart) {
    if (userId == null) {
        return ResponseUtil.unlogin();
    }
    if (cart == null) {
        return ResponseUtil.badArgument();
    }

    Integer productId = cart.getProductId();
    Integer number = cart.getNumber().intValue();
    Integer goodsId = cart.getGoodsId();
    if (!ObjectUtils.allNotNull(productId, number, goodsId)) {
        return ResponseUtil.badArgument();
    }

    //判断商品是否可以购买
    LitemallGoods goods = goodsService.findById(goodsId);
    if (goods == null || !goods.getIsOnSale()) {
        return ResponseUtil.fail(GOODS_UNSHELVE, "商品已下架");
    }

    LitemallGoodsProduct product = productService.findById(productId);
    //判断购物车中是否存在此规格商品
    LitemallCart existCart = cartService.queryExist(goodsId, productId,
                        userId);
    if (existCart == null) {
        //取得规格的信息,判断规格库存
        if (product == null || number > product.getNumber()) {
            return ResponseUtil.fail(GOODS_NO_STOCK, "库存不足");
        }

        cart.setId(null);
        cart.setGoodsSn(goods.getGoodsSn());
```

```
                    cart.setGoodsName((goods.getName()));
                    cart.setPicUrl(goods.getPicUrl());
                    cart.setPrice(product.getPrice());
                    cart.setSpecifications(product.getSpecifications());
                    cart.setUserId(userId);
                    cart.setChecked(true);
                    cartService.add(cart);
                } else {
                    //取得规格的信息，判断规格库存
                    int num = existCart.getNumber() + number;
                    if (num > product.getNumber()) {
                        return ResponseUtil.fail(GOODS_NO_STOCK, "库存不足");
                    }
                    existCart.setNumber((short) num);
                    if (cartService.updateById(existCart) == 0) {
                        return ResponseUtil.updatedDataFailed();
                    }
                }
                return goodscount(userId);
            }
```

2.6 本地测试

本地测试是指开发人员在本地计算机的开发环境中测试项目程序，本节将详细讲解在本地测试本商城系统所有模块的过程。

2.6.1 创建数据库

扫码看视频

本系统使用 MySQL 数据库存储数据，数据库文件存放在 litemall-db\sql 文件夹中，其中，文件 litemall_schema.sql 用于创建数据库和用户权限，文件 litemall_table.sql 用于创建表，文件 litemall_data.sql 用于创建测试数据。

> **注意**：建议使用命令行、MySQL Workbench 或 AppServ 进行导入，如果使用 navicat 导入，可能失败。

将 litemall-db\sql 文件夹中的 SQL 数据文件导入本地 MySQL 数据库后，在 litemall-db 模块的 application-db.yml 文件中配置连接参数和 druid，主要代码如下。

```
spring:
  datasource:
    druid:
```

```
    url: jdbc:mysql://localhost:3306/litemall?useUnicode=true&characterEncoding=
        UTF-8&serverTimezone=UTC&allowPublicKeyRetrieval=true&verifyServer
        Certificate=false&useSSL=false
    driver-class-name: com.mysql.jdbc.Driver
    username: litemall
    password: litemall123456
    initial-size: 10
    max-active: 50
    min-idle: 10
    max-wait: 60000
    pool-prepared-statements: true
    max-pool-prepared-statement-per-connection-size: 20
    validation-query: SELECT 1 FROM DUAL
    test-on-borrow: false
    test-on-return: false
    test-while-idle: true
    time-between-eviction-runs-millis: 60000
    filters: stat,wall
```

在上述代码中需要注意 username 和 password，这两个参数分别代表连接 MySQL 数据库的用户名和密码。

2.6.2 运行后台管理系统

本项目的后台管理系统由 litemall-admin-api 模块和 litemall-admin 模块组成，在运行后台管理系统之前需要先使用 IntelliJ IDEA 运行后端模块 litemall-all，然后使用 Vue 运行前端模块 litemall-admin。

（1）在 IntelliJ IDEA 中找到文件 litemall\litemall-all\src\main\java\org\linlinjava\litemall\Application.java，然后右击此文件，在弹出的快捷菜单中选择 Run 'Application' 命令即可运行后端模块 litemall-all，如图 2-2 所示。

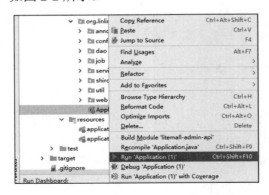

图 2-2　运行后端模块 litemall-all

(2) 使用 Vue 运行前端模块 litemall-admin。首先打开命令行界面，然后输入下面的命令，启动 npm。

```
npm install -g cnpm --registry=https://registry.npm.taobao.org
cd litemall/litemall-admin
cnpm install
cnpm run dev
```

命令运行成功后，会在 npm 界面显示 URL 网址，如图 2-3 所示。

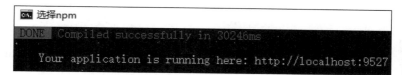

图 2-3　npm 界面

注意：在运行后台模块之前，一定要确保以下两个文件中的 port 一致。

```
D:\litemall\litemall-admin-api\src\main\resources\application.yml
D:\litemall\litemall-admin\config\dep.env.js
```

(3) 在浏览器地址栏中输入 http://localhost:9527 后即可运行后台模块。首先显示登录界面，如图 2-4 所示。

图 2-4　后台登录界面

根据提示输入登录信息后来到后台界面，例如商品列表界面，如图 2-5 所示。

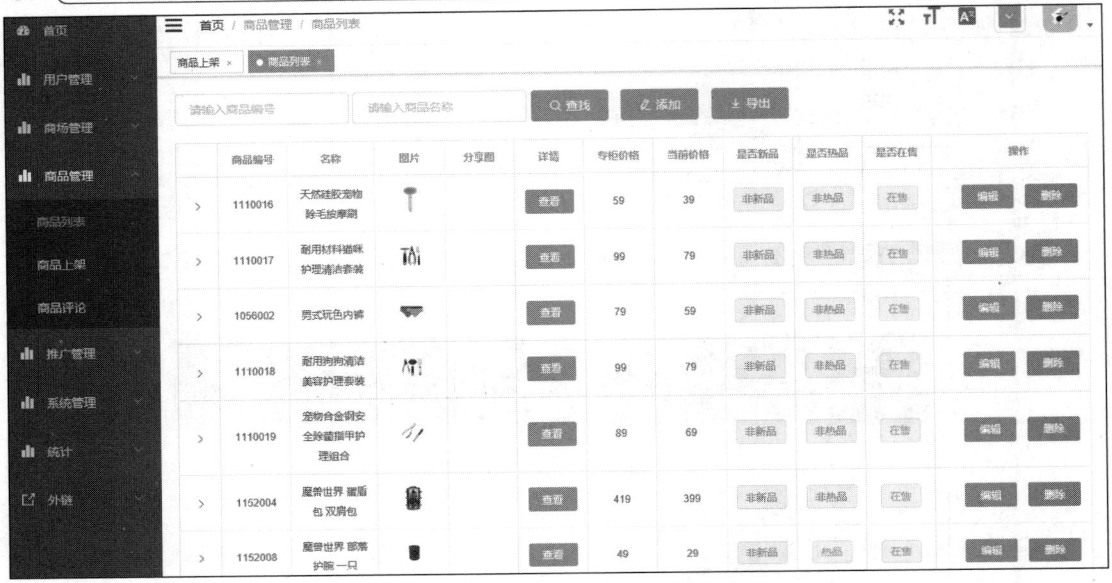

图 2-5　商品列表界面

2.6.3　运行微信小商城子系统

本项目的微信小商城子系统由 litemall-wx-api、litemall-wx 和 renard-wx 共 3 个模块组成。其中，litemall-wx-api 是基于 Spring Boot 技术实现的后端模块，litemall-wx 和 renard-wx 是使用微信小程序实现的前端模块。在调试时，只需运行一个前端模块即可，下面以运行 litemall-wx 模块为例进行讲解。

(1) 在 IntelliJ IDEA 中按照 2.6.2 节中的方法运行 litemall-all 模块，如果已经运行过，则无须重复这个步骤。

(2) 运行微信小商城子系统的前端，登录腾讯微信小程序官方网站，下载并安装微信 Web 开发工具，然后打开微信 Web 开发工具，如图 2-6 所示。

(3) 单击中间的"+"按钮，在弹出的界面中将 litemall-wx 目录下的源码导入微信 Web 开发工具。因为是本地测试，可以单击"测试号"链接，使用微信官方提供的测试号，如图 2-7 所示。

(4) 打开文件 litemall\litemall-wx\config\api.js，确认变量 WxApiRoot 的端口号和后台系统的端口号一致。

(5) 在资源文件 litemall-core\src\main\resources\application-core.yml 中设置 AppID 和密钥，可以使用微信官方提供的测试号。

```
litemall
    wx
        app-id: 开发者申请的 app-id 或测试号
        app-secret: 开发者申请的 app-secret 或测试号
```

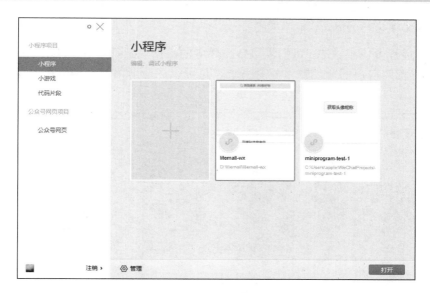

图 2-6　微信 Web 开发工具

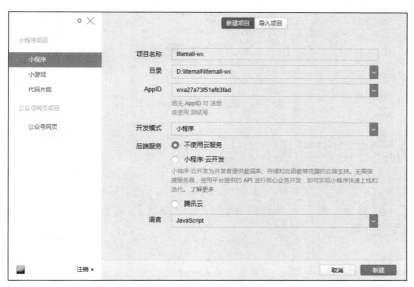

图 2-7　导入微信 Web 开发工具

(6) 在文件 litemall-wx\project.config.json 中设置 AppID；可以使用微信官方提供的测试号。

> 注意：建议开发者关闭当前项目或者直接关闭微信开发者工具，再重新打开(因为此时 litemall-wx 模块的 appid 可能未更新)。

(7) 编译运行微信小程序，可以获取数据库中的数据，在商城中会显示数据库的商品信息。我们可以使用自己的微信账号登录系统，也可以注册新用户。微信小商城模块界面如图 2-8 所示。

(a) 商城首页

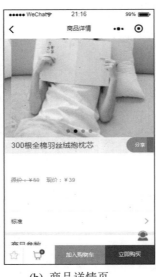
(b) 商品详情页

(c) 购物车页面

图 2-8 微信小商城模块界面

2.7 线上发布和部署

如果读者申请了微信开发者账号并开通了服务号，就可以线上发布自己的商城系统。在开通服务号时，需要通过微信官方审核认证(收费认证)。下面简单介绍线上发布本系统的具体过程。

扫码看视频

2.7.1 微信登录配置

在本系统中有两个地方需要配置微信登录功能，首先是小商城前端 litemall-wx 模块(或

renard-wx 模块)中 project.config.json 文件的 appid，其次是小商城后端 litemall-core 模块的 application-core.yml 文件。

```
litemall:
  wx:
    app-id: 申请的账号
    app-secret: 申请的密码
```

这里的 app-id 和 app-secret 需要开发者在微信公众平台注册获取，而不能使用测试号。

2.7.2 微信支付配置

在 litemall-core 模块的 application-core.yml 文件中配置微信支付信息，主要代码如下。

```
litemall:
  wx:
    mch-id: 111111
    mch-key: xxxxxx
    notify-url: https://www.example.com/wx/order/pay-notify
```

参数说明如下。

(1) mch-id 和 mch-key：需要开发者在微信商户平台注册获取。

(2) notify-url：项目上线以后微信支付回调地址。当微信支付成功或者失败时，微信商户平台将向回调地址发送成功或者失败的数据，因此需要确保该地址是 litemall-wx-api 模块的 WxOrderController 类的 payNotify 方法所服务的 API 地址。

> **注意**：在开发阶段，可以采用一些技术实现临时外网地址映射本地功能，开发者可以在网络上搜索关键字"微信内网穿透"自行学习。

2.7.3 配置邮件通知

邮件通知是指在用户下单后，系统会自动向 sendto 用户发送一封邮件，告知用户下单的订单信息。当然，如果不需要邮件通知订单信息，可以默认关闭。在 litemall-core 模块的文件 application-core.yml 中可配置邮件通知服务，主要代码如下。

```
litemall:
  notify:
    mail:
      # 邮件通知配置，邮箱一般用于接收业务通知，例如收到新的订单，sendto 用于定义邮件接收者，
      # 通常为商城运营人员
      enable: false
      host: smtp.exmail.qq.com
```

```
        username: ex@ex.com.cn
        password: XXXXXXXXXXXX
        sendfrom: ex@ex.com.cn
        sendto: ex@qq.com
```

配置邮件通知功能的基本流程如下。

(1) 在邮件服务器开启 SMTP 服务。

(2) 开发者在配置文件中设置 enable 的值为 true，并为其他信息设置相应的值，建议使用 QQ 邮箱。

(3) 当配置好邮箱信息以后，可以运行 litemall-core 模块的 MailTest 测试类进行发送测试，然后登录邮箱查看邮件是否接收成功。

2.7.4 短信通知配置

目前，短信通知场景只支持支付成功、验证码、订单发送、退款成功 4 种情况，以后微信可能会继续扩展新的模块。在 litemall-core 模块的文件 application-core.yml 中可配置短信通知服务，主要代码如下。

```
litemall:
  notify:
    # 短消息模板通知配置
    # 短信息用于通知客户,例如发货短信通知,注意配置格式: template-name,template-templateId
    # 可参考 NotifyType 枚举值
    sms:
      enable: false
      appid: 111111111
      appkey: xxxxxxxxxxxxxx
      template:
      - name: paySucceed
        templateId: 156349
      - name: captcha
        templateId: 156433
      - name: ship
        templateId: 158002
      - name: refund
        templateId: 159447
```

配置短信通知的基本流程如下。

(1) 登录腾讯云短信平台，申请开通短信功能，然后设置 4 个场景的短信模板。

(2) 在配置文件中设置 enable 的值为 true，然后设置其他信息，包括腾讯云短信平台申请的 appid 等值。建议使用腾讯云短信平台，也可以自行测试其他短信云平台。

(3) 当配置好短信通知功能以后，可以通过 litemall-core 模块中的测试 SmsTest 类进行测试，测试时需要设置手机号和模板所需要的参数值。单独启动 SmsTest 测试类发送短信，然后查看手机是否接收成功。

2.7.5 系统部署

读者可以根据自己的实际情况来选择部署方案，下面是较为常用的 4 种部署方案。

(1) 可以在同一云主机中安装一个 Spring Boot 服务，以及 litemall-admin、litemall-admin-api 和 litemall-wx-api 这 3 个服务。

(2) 可以在同一云主机中仅安装一个 tomcat/nginx 服务器，并部署 litemall-admin 静态页面分发服务，然后部署两个 Spring Boot 的后端服务。

(3) 可以把 litemall-admin 静态页面托管给第三方 CDN，然后部署两个后端服务。

(4) 可以部署到多个服务器，然后采用集群式并发服务。

2.7.6 技术支持

本项目的开发团队一直在维护本系统，读者可以登录 GitHub，搜索 litemall 找到本项目，及时了解本项目的更新和升级情况。建议读者通过 releases 模块了解最新的更新信息，也可以在码云找到本项目的升级源码，具体地址是 https://gitee.com/linlinjava/litemall。

另外，开发团队提供了完善的说明文档，地址是 https://linlinjava.gitbook.io/litemall/。调试过程中的常见问题，可在源码文件 litemall/doc/FAQ.md 中查看解决方案。同时，本开发团队还提供了技术支持 QQ 群，具体群号可登录 https://gitee.com/linlinjava/litemall 获取。

2.7.7 项目参考

本项目基于或参考以下开源项目。

(1) nideshop-mini-program：基于 Node.js+MySQL 开发的开源微信小程序商城(微信小程序)。

项目参考：
- litemall 项目数据库基于 nideshop-mini-program 项目数据库。
- litemall 项目的 litemall-wx 模块基于 nideshop-mini-program 开发。

(2) vue-element-admin：一个基于 Vue 和 Element 的后台集成方案。

项目参考：litemall 项目的 litemall-admin 模块的前端框架基于 vue-element-admin 项目修改扩展。

(3) mall-admin-web：mall-admin-web 是一个电商后台管理系统的前端项目，基于

Vue+Element 实现。

项目参考：litemall 项目的 litemall-admin 模块的一些页面布局样式参考了 mall-admin-web 项目。

(4) biu：管理后台项目开发脚手架，基于 vue-element-admin 和 springboot 搭建，以前后端分离方式开发和部署。

项目参考：litemall 项目的权限管理功能参考了 biu 项目。

(5) vant-mobile-mall：基于有赞 vant 组件库的移动商城。

项目参考：litemall 项目的 litemall-vue 模块基于 vant-mobile-mall 项目开发。

第 3 章 图书借阅管理系统

过去人们使用传统的人工方式管理图书馆的日常工作，不足之处显而易见，处理借书、还书业务流程的效率很低。随着计算机的普及，人们已经可以使用软件管理系统提高借阅图书管理的效率。本章将使用 Java 语言开发一个图形化界面的图书借阅管理系统，展示 Java 语言在桌面项目中的应用过程。本章项目由 JavaFX+JFoenix+MySQL 实现。

3.1 背景介绍

图书馆里的书籍种类繁多，图书馆里的图书管理、借阅管理、读者管理等管理工作也非常复杂。随着学校人数的增多，学生对知识需求的增大，图书馆的图书借阅量也大幅上升，这导致学生经常借不到自己想要的书，同时也给图书馆的图书分类及管理增添了很多问题。在 21 世纪这个飞速发展的时代，利用信息化技术提高生产力势在必行。开发一个满足基本的图书借阅和管理需求的智能化系统，实现图书信息的智能化，能大幅减轻图书馆管理人员的工作负担。

扫码看视频

使用图书借阅管理系统可以管理不同种类且数量繁多的图书，提高图书馆图书管理工作的效率，减少工作中可能出现的错误，为借阅者提供更好的服务，它是提高学校自动化水平的重要组成部分。

3.2 系统分析

本节要做好开发图书借阅管理系统的准备工作，内容包括系统需求分析和系统功能分析。

扫码看视频

3.2.1 系统需求分析

要开发一个信息系统，首先要对信息系统的需求进行分析。需求分析要做的工作是深入描述软件的功能和性能，确定实现软件的限制与其他系统元素的接口细节，定义软件的其他有效性需求。

要获得当前系统的处理流程，在此首先假设当前系统是手工处理系统。手工处理系统的流程大致是这样的：读者将要借的书和借阅证交给工作人员，工作人员将每本书附带的描述书信息的书卡和借阅证一起放在一个小格栏，并在借阅证和每本书上贴上借阅信息，这样借书过程就完成了。还书时，读者将要还的图书交给工作人员，工作人员按图书信息找到相应的书卡和借阅证，并填写相应的还书信息。

（1）抽象出当前系统的逻辑模型：在理解当前系统"怎么做"的基础上，抽取其"做什么"的本质，从而从当前系统的物理模型抽象出当前系统的逻辑模型。在物理模型中有许多物理因素，有些非本质的物理因素会成为不必要的负担，因而需要对物理模型进行分析，区分出本质的因素和非本质的因素，去掉那些非本质的因素即可获得反映系统本质的逻辑模型。

(2) 建立目标系统的逻辑模型：分析目标系统与当前系统逻辑上的差别，明确目标系统到底要"做什么"，从而从当前系统的逻辑模型导出目标系统的逻辑模型。

通过上述流程，我们对新的图书处理流程进行整理，总结图书馆借还书的过程如下。

- 借书过程：读者从书架上选到所需图书后，将图书和借阅证交管理人员，管理人员用码阅读器将图书条码和借阅证上的读者条码读入处理系统。系统根据读者条码从读者文件和借阅文件中找到相应记录；根据图书条码从图书文件中找到相应记录，读者如果有下列情况之一将不予办理借书手续：超期未还，未交罚款。

- 还书过程：还书时，读者只要将书交给管理人员，管理员将书上的图书条码读入系统，系统从借阅文件上找到相应记录，填上还书日期后写入借阅历史文件，并从借阅文件上删去相应记录；同时系统对借还书日期进行计算并判断是否超期，若不超期则结束过程，若超期则计算出超期天数、罚款数并打印罚款通知单，记入罚款文件，同时在读者借阅文件上做止借标记。当读者交来罚款收据后，系统根据读者条码查罚款文件，将相应记录写入罚款历史文件，并从罚款文件上删除该记录，同时去掉读者借阅文件中的止借标记。

3.2.2 系统功能分析

图书馆管理系统的用户主要是各个学校的图书馆，具体功能如下。

1) 基本信息管理模块

基本信息管理模块的主要功能包括读者信息管理、图书类别管理、图书信息管理(包括添加、删除和修改)。

2) 图书借阅(借书、还书)管理模块

针对图书馆最主要的借书、还书，图书查找，借阅超期查看等需求，各功能模块完成相关数据的记录。

3) 用户信息管理模块

用户信息管理模块的功能比较简单。用户分为管理员用户和普通用户。系统管理员用户可以创建用户、修改用户信息以及删除用户，普通用户只能够修改自己的用户信息。

根据需求分析中用户的要求，设计系统的体系结构，如图 3-1 所示。图中的每一个结点都是一个最小的功能模块。每一个功能模块都需要针对不同的数据库表完成相同的数据库操作，即添加记录、删除记录、查询记录、更新记录。

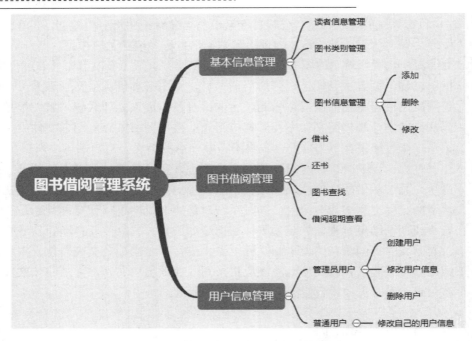

图 3-1　图书借阅管理系统功能模块示意图

3.3　数据库设计

本项目的开发主要包括后台数据库的建立、测试数据的录入以及前台应用程序的开发三个方面。数据库设计是系统设计开发的一个重要组成部分，数据库设计得好坏会直接影响程序编码的复杂程度。

扫码看视频

3.3.1　选择数据库

开发数据库管理信息系统时，要根据用户需求、系统功能和性能要求等因素，选择后台数据库和相应的数据库访问接口。后台数据库的选择需要考虑用户需求、系统功能和性能要求等因素。考虑到本系统所要管理的数据量比较大，且需要多用户同时访问，本项目将使用 MySQL 作为后台数据库管理平台。

3.3.2　数据库结构的设计

由需求分析的规划可知，整个项目对象有 5 种信息，所以对应的数据库也需要包含这 5

种信息，从而系统需要包含 5 个数据库表，分别如下。

- book：图书信息表。
- book_type：图书类型表。
- borrow：图书借阅信息表。
- reader：读者用户信息表。
- user：系统用户信息表。

下面将详细讲解各个数据库表的具体结构信息。

(1) 图书信息表 book 用来保存图书信息，结构如表 3-1 所示。

表 3-1　图书信息表的结构

名　称	类　型	空	默 认 值	备　注
Id	varchar(32)	否		图书编号
name	varchar(100)	是	<空>	图书名称
type	int(11)	是	<空>	图书类别的编号
typestr	varchar(50)	是	<空>	图书类别的名称
author	varchar(50)	是	<空>	作者
translator	varchar(50)	是	<空>	译者
publisher	varchar(1024)	是	<空>	出版社
publish_time	date	是	<空>	出版时间
stock	int(11)	是	<空>	库存容量
price	numeric(5,2)	是	<空>	价格

(2) 图书类型表(book_type)用来保存图书类型信息，表结构如表 3-2 所示。

表 3-2　图书类型表结构

名　称	类　型	空	默 认 值	备　注
Id	int(11)	否	<auto_increment>	编号
type	varchar(20)	是	<空>	图书类别

(3) 图书借阅信息表(borrow)用来保存图书借阅情况信息，表结构如表 3-3 所示。

(4) 读者用户信息表(reader)用来保存借阅图书的读者信息，表结构如表 3-4 所示。

(5) 系统用户信息表(user)用来保存系统内的所有用户信息，包括学生和管理员，表结构如表 3-5 所示。

表 3-3 图书借阅信息表结构

名称	类型	空	默认值	备注
Id	int(11)	否	<auto_increment>	借阅流水号
book_id	varchar(50)	是	<空>	图书编号
reader_id	varchar(50)	是	<空>	读者编号
borrow_date	date	是	<空>	借出时间
back_date	date	是	<空>	到期时间
is_back	smallint(1)	是	<空>	是否归还

表 3-4 读者用户信息表结构

名称	类型	空	默认值	备注
Id	varchar(32)	否		读者编号
name	varchar(50)	是	<空>	读者名称
pass	varchar(50)	是	<空>	读者密码
type	varchar(20)	是	<空>	读者类型
sex	char(1)	是	<空>	性别
max_num	int(3)	是	<空>	最大可借数
days_num	int(11)	是	<空>	可借天数
forfeit	numeric(5,2)	是	<空>	罚款金额

表 3-5 系统用户信息表结构

名称	类型	空	默认值	备注
Id	varchar(32)	否		用户编号
name	varchar(50)	否		用户名称
pass	varchar(50)	是	<空>	用户密码
email	varchar(20)	否		用户邮箱
is_admin	smallint(1)	是	<空>	是否为管理员

3.4 系统框架设计

系统框架设计步骤是整个项目的基础,此过程需要经过以下 4 个阶段。
(1) 开发环境:Windows 11 操作系统。
(2) 数据库:MySQL。
(3) 开发工具:IntelliJ IDEA。
(4) 导入引用包:第三方框架 JFoenix。

扫码看视频

3.4.1 创建工程

打开 IntelliJ IDEA,新建一个 Java 工程,工程的目录结构如图 3-2 所示。在新建的工程下自动生成 src 目录,用于存放源代码。

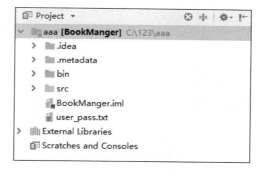

图 3-2 工程的目录结构

3.4.2 导入引用包

(1) 因为本实例使用第三方框架 JFoenix 以提高程序界面的美观性,所以需要在其官网下载 jar 包。开源框架 JFoenix 的官网主页是 www.jfoenix.com,如图 3-3 所示。
(2) 单击 GitHub 按钮,进入 JFoenix 源码的托管网站,如图 3-4 所示。
(3) 往下滚动页面,单击 download jar 链接下载适合自己版本的 jar 包,如图 3-5 所示。本书下载的是 jfoenix-8.0.4.jar。
(4) 在前面创建的 Java 工程的主目录中新建一个名为 jar 的包,将刚下载的文件 jfoenix-8.0.4.jar 复制到这个包中。因为本项目还用到了 MySQL 数据库,所以将 JDBC MySQL 包也一块复制到这个包目录中,如图 3-6 所示。

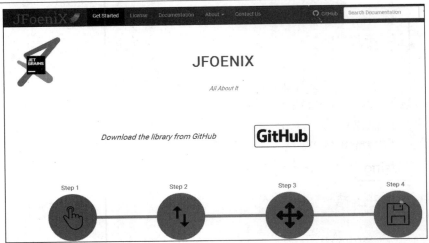

图 3-3　JFoenix 官网

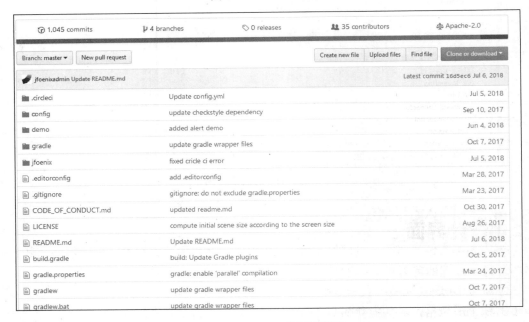

图 3-4　JFoenix 的托管网站

(5) 右击 jfoenix-8.0.4.jar，在弹出的快捷菜单中选择 Add as Library 命令，将此包添加到当前 Java 工程中，如图 3-7 所示。用同样的方法，将 MySQL 数据库连接包 mysql-connector-java-5.1.7-bin.jar 也添加到当前 Java 工程中。

第 3 章 图书借阅管理系统

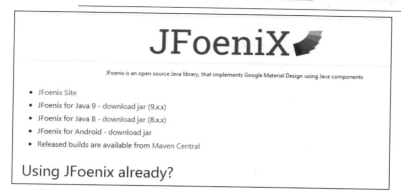

图 3-5 下载适合自己版本的 jar 包

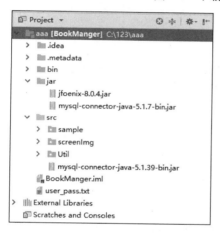

图 3-6 复制引用包

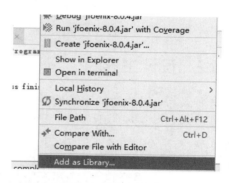

图 3-7 添加引用包

3.5 设计界面

为了提高项目的美观性，可以为界面设计一个漂亮的背景。本项目采用的第三方框架 JFoenix 是基于 JavaFX 的一个框架，使用了 Android 设计的 MaterialDesighn 风格，界面控件的外观十分漂亮。

扫码看视频

3.5.1 使用 JavaFX Scene Builder 设计界面

因为 JFoenix 是基于 JavaFX 的一个框架，所以我们使用设计工具 JavaFX Scene Builder 来设计界面。在使用 JavaFX Scene Builder 之前，需要先将前面下载的包 jfoenix-8.0.4.jar 导

107

入 JavaFX Scene Builder，如图 3-8 所示。

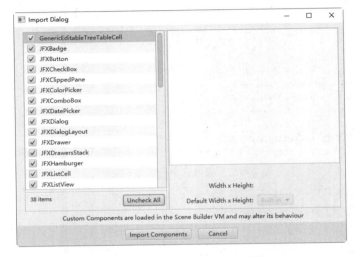

图 3-8　在 JavaFX Scene Builder 中导入 JFoenix

3.5.2　设计主界面

用户登录后，首先进入系统的主界面。主界面包括：①位于界面顶部的菜单栏，它将系统所具有的功能进行归类展示；②位于菜单栏下面的工具栏，它将系统的常用功能以按钮的方式展示；③位于工具栏下面的工作区，它是进行各种功能操作的区域，如图 3-9 所示。

图 3-9　主界面

因为主界面是整个系统通往各个功能模块的窗口,所以要将各个功能模块的窗体加入主界面中。同时要保证各窗体在主界面中布局合理,让用户方便操作,因此在主窗体中应加入整个系统的入口函数 main,通过执行该方法进而执行整个系统。主窗体类 Main.java 的具体实现代码如下。

```java
public class Main extends Application {

    private Stage mainStage;

    @Override
    public void start(Stage primaryStage) throws Exception{
        mainStage = primaryStage;
        mainStage.setResizable(false);
        //设置窗口的图标
        mainStage.getIcons().add(new Image(
                Main.class.getResourceAsStream("logo.png")));
        FXMLLoader loader = new FXMLLoader(getClass().getResource("sample.fxml"));
        Parent root = loader.load();
        primaryStage.setTitle("图书管理系统");
        Controller controller = loader.getController();
        controller.setApp(this);
        Scene scene = new Scene(root, 700, 460);
        scene.getStylesheets().add(Main.class.getResource ("main.css").toExternalForm());
        primaryStage.setScene(scene);
        primaryStage.show();
    }

    public void gotoMainUi(String userId) {
        try {
            FXMLLoader loader = new FXMLLoader(getClass().getResource("main_ui.fxml"));
            Parent root = loader.load();
            mainStage.setTitle("图书管理系统");
            MainUiController controller = loader.getController();
            controller.setApp(this);
            controller.setMyName(userId);
            Scene scene = new Scene(root, 700, 500);
            scene.getStylesheets().add(Main.class.getResource
                    ("main.css").toExternalForm());
            mainStage.setScene(scene);
            mainStage.show();
        } catch (Exception e) {
            System.out.println(e.getMessage());
        }

    }
```

```java
public void gotoReaderUi(String id) {
    try {
        FXMLLoader loader = new FXMLLoader(getClass().getResource("reader_ui.fxml"));
        Parent root = loader.load();
        mainStage.setTitle("图书管理系统");
        ReaderUi controller = loader.getController();
        controller.setApp(this);
        controller.setUserInfo(id);
        Scene scene = new Scene(root, 700, 460);
        scene.getStylesheets().add(Main.class.getResource
                ("main.css").toExternalForm());
        mainStage.setScene(scene);
        mainStage.show();
    } catch (Exception e) {
        System.out.println(e.getMessage()+e.toString());
    }

}

public void closeWindow() {
    mainStage.close();
}

public void hideWindow(){ mainStage.hide();}

public void showWindow(){ mainStage.show();}


public static void main(String[] args) {
    launch(args);
}

public void gotoLoginUi() {
    try {
        FXMLLoader loader = new FXMLLoader(getClass().getResource("sample.fxml"));
        Parent root = loader.load();
        mainStage.setTitle("图书管理系统");
        Controller controller = loader.getController();
        controller.setApp(this);
        Scene scene = new Scene(root, 700, 460);
        scene.getStylesheets().add(Main.class.getResource
                ("main.css").toExternalForm());
        mainStage.setScene(scene);
        mainStage.show();
    } catch (Exception e) {
```

```
            System.out.println(e.toString());
        }
    }
}
```

3.6 为数据库表添加对应的类

类是面向对象编程的核心。为了便于对数据库进行控制,需要为项目中的每一个数据库表创建一个独立的类,类的成员变量对应数据库中表的列,成员函数对应成员变量和对表的操作,这样可以灵活控制每个数据库表。为了提高代码的重用性,需从多个表中取出信息并组合成一个对象,用于在整个流程中进行访问。

扫码看视频

3.6.1 Book 类

类 Book 用于对数据库中表 Book 进行操作,具体实现代码如下。

```
public class Book {
    private String Id;
    private String name;
    private int type;
    private String typeStr;
    private String author;
    private String translator;
    private String publisher;
    private String publishTime;
    private int stock;
    private double price;

    public Book(String id, String name, int type, String typeStr, String author, String translator, String publisher, String publishTime, int stock, double price) {
        this.Id = id;
        this.name = name;
        this.type = type;
        this.typeStr = typeStr;
        this.author = author;
        this.translator = translator;
        this.publisher = publisher;
        this.publishTime = publishTime;
```

```java
        this.stock = stock;
        this.price = price;
    }

    public Book() {

    }

    public String getId() {
        return Id;
    }

    public void setId(String id) {
        Id = id;
    }

    public String getName() {
        return name;
    }

    public void setName(String name) {
        this.name = name;
    }

    public int getType() {
        return type;
    }

    public void setType(int type) {
        this.type = type;
    }

    public String getAuthor() {
        return author;
    }

    public void setAuthor(String author) {
        this.author = author;
    }

    public String getTranslator() {
        return translator;
    }

    public void setTranslator(String translator) {
        this.translator = translator;
    }
```

```java
    public String getPublisher() {
        return publisher;
    }

    public void setPublisher(String publisher) {
        this.publisher = publisher;
    }

    public String getPublishTime() {
        return publishTime;
    }

    public void setPublishTime(String publishTime) {
        this.publishTime = publishTime;
    }

    public int getStock() {
        return stock;
    }

    public void setStock(int stock) {
        this.stock = stock;
    }

    public double getPrice() {
        return price;
    }

    public void setPrice(double price) {
        this.price = price;
    }

    public String getTypeStr() {
        return typeStr;
    }

    public void setTypeStr(String typeStr) {
        this.typeStr = typeStr;
    }
}
```

为了提高开发效率，读者不必手写每个字段对应的 get、set 方法，而是在定义 Book 类后，在类中直接添加成员变量。方法是选中要创建的 Getter 和 Setter 代码字段，如图 3-10 所示。然后按 Alt+Insert 组合键，在弹出的窗口中根据自己的需要选择 Getter 或 Setter，如

图3-11所示。

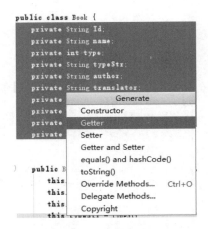

图 3-10 选中代码　　　　图 3-11 选择 Getter

3.6.2 借阅类 Borrow

类 Borrow 用于对数据库中表 borrow 进行定义，具体实现代码如下。

```java
public class Borrow {
    private String id;
    private String bookId;
    private String readerId;
    private String borrowDate;
    private String backDate;
    private int isBack;

    public Borrow(String id, String bookId, String readerId, String borrowDate,
                  String backDate, int isBack) {
        this.id = id;
        this.bookId = bookId;
        this.readerId = readerId;
        this.borrowDate = borrowDate;
        this.backDate = backDate;
        this.isBack = isBack;
    }

    public Borrow() {

    }

    public String getId() {
```

```java
        return id;
    }

    public void setId(String id) {
        this.id = id;
    }

    public String getBookId() {
        return bookId;
    }

    public void setBookId(String bookId) {
        this.bookId = bookId;
    }

    public String getReaderId() {
        return readerId;
    }

    public void setReaderId(String readerId) {
        this.readerId = readerId;
    }

    public String getBorrowDate() {
        return borrowDate;
    }

    public void setBorrowDate(String borrowDate) {
        this.borrowDate = borrowDate;
    }

    public String getBackDate() {
        return backDate;
    }

    public void setBackDate(String backDate) {
        this.backDate = backDate;
    }

    public int getIsBack() {
        return isBack;
    }

    public void setIsBack(int isBack) {
        this.isBack = isBack;
    }
}
```

> **注意**：类 Reader 用于对数据库中表 reader 进行操作，类 User 用于对数据库中表 user 进行操作，为节省篇幅，不再列出这两个实体类的具体实现代码。

3.7 系统登录模块

为了增强系统的安全性，需要设置只有通过系统身份验证的用户才能够使用本系统，为此必须增加一个系统登录模块。

扫码看视频

3.7.1 登录验证

通过文件 Controller.java 实现登录验证功能，以确保只有合法的用户才能登录。文件 Controller.java 的主要实现代码如下：

```java
public void initialize(URL location, ResourceBundle resources) {
    rememberInfo.setSelected(true);
    RequiredFieldValidator validator = new RequiredFieldValidator();
    validator.setMessage("请输入用户名...");
    tf_user.getValidators().add(validator);
    tf_user.focusedProperty().addListener((o,oldVal,newVal)->{
        if(!newVal) tf_user.validate();
    });
    RequiredFieldValidator validator2 = new RequiredFieldValidator();
    validator2.setMessage("请输入密码...");
    tf_passWord.getValidators().add(validator2);
    tf_passWord.focusedProperty().addListener((o,oldVal,newVal)->{
        if(!newVal) tf_passWord.validate();
    });
    rb_duzhe.setSelected(true);
    prgs_login.setVisible(false);
    String str = FileUtil.getUserAndPass();
    Pattern p = Pattern.compile("[#]+");
    String[] result = p.split(str);
    if (result.length >= 1) {
        tf_user.setText(result[0]);
    }
    if (result.length >= 2) {
        tf_passWord.setText(result[1]);
    }
}
/**
 * 登录按钮单击事件
```

```java
     */
    @FXML
    public void onStart() {
        System.out.println("ok");
        prgs_login.setVisible(true);
        //创建线程登录
        myProgress myProgress = new myProgress(prgs_login);
        thread = new Thread(myProgress);
        thread.setPriority(Thread.MAX_PRIORITY);
        thread.start();
        if (rememberInfo.isSelected()) {
            FileUtil.setUserAndPass(tf_user.getText(), tf_passWord.getText());
        }else{
            FileUtil.setUserAndPass(tf_user.getText(), "");
        }

        //登录界面控件不可见
        setDisable(true);
    }
    /**
     * 登录期间------组件的控制-----登录界面控件不可见
     */
    public void setDisable(Boolean bool) {
        btn_start.setDisable(bool);
        tf_user.setDisable(bool);
        tf_passWord.setDisable(bool);
        rememberInfo.setDisable(bool);
    }
    /**
     * 检查并登录
     */
    private void doCheckUser() {
        if (identity.getSelectedToggle() == rb_duzhe) {
            if (DataBaseUtil.checkReader(tf_user.getText().trim(),
                                    tf_passWord.getText())) {
                myApp.gotoReaderUi(tf_user.getText());
            } else {
                setDisable(false);
                Alert alert = new Alert(Alert.AlertType.CONFIRMATION);
                alert.setAlertType(Alert.AlertType.ERROR);
                alert.setTitle("登录失败！");
                alert.show();
            }
        } else if (identity.getSelectedToggle() == rb_gzry) {
```

```java
            if (DataBaseUtil.checkUser(tf_user.getText().trim(),
                            tf_passWord.getText())) {
                myApp.gotoMainUi(tf_user.getText());
            } else {
                setDisable(false);
                Alert alert = new Alert(Alert.AlertType.CONFIRMATION);
                alert.setAlertType(Alert.AlertType.ERROR);
                alert.setTitle("登录失败!");
                alert.show();
            }
        }
    }
    /**
     * 忘记密码
     */
    @FXML
    public void forgotPass() {
        myApp.hideWindow();
        Stage myStage=new Stage();
        myStage.setResizable(false);
        //设置窗口的图标
        myStage.getIcons().add(new Image(
            Main.class.getResourceAsStream("logo.png")));
        FXMLLoader loader = new FXMLLoader(getClass().getResource("forgotPass.fxml"));
        Parent root = null;
        try {
            root = loader.load();
        } catch (IOException e) {
            e.printStackTrace();
        }
        ForgotPass con = loader.getController();
        con.setMyApp(myApp);
        con.setController(myStage);
        myStage.setTitle("忘记密码");
        Scene scene = new Scene(root, 475, 400);
        scene.getStylesheets().add(Main.class.getResource ("main.css").toExternalForm());
        myStage.setScene(scene);
        myStage.show();
        
    }
    
    @FXML
    public void logUp() {
        myApp.hideWindow();
        Stage myLogupStage=new Stage();
```

```java
    myLogupStage.setResizable(false);
    //设置窗口的图标
    myLogupStage.getIcons().add(new Image(
            Main.class.getResourceAsStream("logo.png")));
    FXMLLoader loader = new FXMLLoader(getClass().getResource("user_logUp.fxml"));
    Parent root = null;
    try {
        root = loader.load();
    } catch (IOException e) {
        e.printStackTrace();
    }
    UserLogUp con = loader.getController();
    con.setMyApp(myApp);
    con.setController(myLogupStage);
    myLogupStage.setTitle("注册");
    Scene scene = new Scene(root, 475, 400);
    scene.getStylesheets().add(Main.class.getResource ("main.css").toExternalForm());
    myLogupStage.setScene(scene);
    myLogupStage.show();
}
/**
 * 登录界面—单击登录按钮后---启用新的线程检查用户身份是否正确
 */
class myProgress implements Runnable {
    private JFXProgressBar prgs_login;
    myProgress(JFXProgressBar prgs_login) {
        this.prgs_login = prgs_login;
    }
    @Override
    public void run() {
        try {
            for (int i = 0; i <= 100; i++) {
                prgs_login.setProgress(i);
            }
            sleep(100);
            //更新 JavaFX 的主线程的代码放在此处
            Platform.runLater(Controller.this::doCheckUser);
        } catch (Exception ignored) {
        }
    }
}
```

登录验证界面的执行效果如图 3-12 所示。

图 3-12 登录验证界面

3.7.2 忘记密码

通过文件 ForgotPass.java 实现修改密码功能，以帮助忘记密码的用户修改密码。文件 ForgotPass.java 的主要实现代码如下。

```java
public void initialize(URL location, ResourceBundle resources) {
    validator = new RequiredFieldValidator();
    validator.setMessage("请输入有效的邮箱...");
    tf_frg_email.getValidators().add(validator);
    tf_frg_email.focusedProperty().addListener((o,oldVal,newVal)->{
        if(!newVal) tf_frg_email.validate();
    });

    validator2 = new RequiredFieldValidator();
    validator2.setMessage("请输入有效的密码...");
    tf_frg_newPass.getValidators().add(validator2);
    tf_frg_newPass.focusedProperty().addListener((o,oldVal,newVal)->{
        if(!newVal) tf_frg_newPass.validate();
    });
}

public void setMyApp(Main myApp) {
    this.myApp = myApp;
}

public void setController(Stage myStage) {
    this.myStage = myStage;
}

@FXML
```

```java
public void confirm() {
    boolean email_ready = false;
    boolean pass_ready = false;
    String email = tf_frg_email.getText().trim();
    String newPass = tf_frg_newPass.getText().trim();
    if (!email.equals("") || !newPass.equals("")) {
        if (email.endsWith(".com") || email.contains("@")) {
            email_ready = true;
        } else {
            tf_frg_email.setText("");
            tf_frg_email.validate();
        }
        if (newPass.length() >= 8) {
            pass_ready = true;
        } else {
            tf_frg_newPass.setText("");
            tf_frg_newPass.validate();
        }
        if (pass_ready && email_ready) {
            boolean isok;
            if (rdb_reader.isSelected()) {
                isok = DataBaseUtil.alterReaderPass(email, newPass);
            } else {
                isok = DataBaseUtil.alterUserPass(email, newPass);
            }
            if (isok) {
                Alert alert = new Alert(Alert.AlertType.CONFIRMATION);
                alert.setAlertType(Alert.AlertType.INFORMATION);
                alert.setTitle("修改成功！");
                alert.showAndWait();
                myApp.showWindow();
                myStage.close();
            }
        }

    } else {
        tf_frg_email.setText("");
        tf_frg_newPass.setText("");
        tf_frg_email.validate();
        tf_frg_newPass.validate();
        return;
    }
    //myApp.showWindow();
    //myStage.close();
}

@FXML
public void logUp() {
    Stage myLogupStage=new Stage();
```

```
myLogupStage.setResizable(false);
//设置窗口的图标
myLogupStage.getIcons().add(new Image(
        Main.class.getResourceAsStream("logo.png")));
FXMLLoader loader = new FXMLLoader(getClass().getResource("user_logUp.fxml"));
Parent root = null;
try {
    root = loader.load();
} catch (IOException e) {
    e.printStackTrace();
}
UserLogUp con = loader.getController();
con.setMyApp(myApp);
con.setController(myLogupStage);
myLogupStage.setTitle("注册");
Scene scene = new Scene(root, 475, 400);
scene.getStylesheets().add(Main.class.getResource ("main.css").toExternalForm());
myLogupStage.setScene(scene);
myLogupStage.show();
myStage.close();
}
```

忘记密码界面的执行效果如图 3-13 所示。

图 3-13　忘记密码界面

3.7.3　新用户注册

通过文件 UserLogUp.java 实现新用户的注册功能，若注册成功，用户信息将被添加到系统数据库中。文件 UserLogUp.java 的主要实现代码如下。

```java
public void initialize(URL location, ResourceBundle resources) {
    validator = new RequiredFieldValidator();
    validator.setMessage("请输入...");
    tf_id.getValidators().add(validator);
    tf_id.focusedProperty().addListener((o,oldVal,newVal)->{
       if(!newVal) tf_id.validate();
    });

    tf_userName.getValidators().add(validator);
    tf_userName.focusedProperty().addListener((o,oldVal,newVal)->{
       if(!newVal) tf_userName.validate();
    });

    tf_PassWord.getValidators().add(validator);
    tf_PassWord.focusedProperty().addListener((o,oldVal,newVal)->{
       if(!newVal) tf_PassWord.validate();
    });

    tf_email.getValidators().add(validator);
    tf_email.focusedProperty().addListener((o,oldVal,newVal)->{
       if(!newVal) tf_email.validate();
    });

}

public void setMyApp(Main myApp) {
    this.myApp = myApp;
}

public void setController(Stage myStage) {
    this.myStage = myStage;
}

@FXML
public void goBackLogin() {
    myApp.showWindow();
    myStage.close();
}

@FXML
public void confirm() {
    String userName = tf_userName.getText().trim();
    String passWord = tf_PassWord.getText().trim();
    String email = tf_email.getText().trim();
    String id = tf_id.getText().trim();
    String sex = "";
    if (rb_sex_man.isSelected()) {
```

```
            sex = "男";
        }else{
            sex = "女";
        }
        if (!id.equals("") || !email.equals("") || !passWord.equals("")
|| !userName.equals("")) {
            boolean isok = false;
            if (rdb_reader.isSelected()) {
                Reader reader = new Reader(id,userName,passWord,"学生",sex,12,30,0);
                isok = DataBaseUtil.addNewReader(reader);
            } else {
                User user = new User(id,userName, passWord, email,1);
                isok = DataBaseUtil.addNewUser(user);
            }
            if (isok) {
                Alert alert = new Alert(Alert.AlertType.CONFIRMATION);
                alert.setAlertType(Alert.AlertType.INFORMATION);
                alert.setTitle("注册成功! ");
                alert.showAndWait();
                myApp.showWindow();
                myStage.close();
            }
        } else {

            return;
        }
    }
```

新用户注册界面的执行效果如图 3-14 所示。

图 3-14　新用户注册界面

3.8 基本信息管理模块

在本项目中,基本信息管理模块是指对以下信息进行管理。
- 读者信息。
- 图书类别信息。
- 图书信息。
- 借书处理。
- 还书处理。

扫码看视频

上述基本操作的功能都是在文件 MainUiController.java 中实现的,接下来将详细讲解上述功能的实现过程。

3.8.1 读者信息管理

读者信息管理功能模块包括读者信息添加、修改和删除三个部分。在文件 MainUiController.java 中编写了对应函数,分别用于实现这三个功能。

1) 初始化界面

通过函数 initReaderAddUi()默认显示读者初始化界面,具体实现代码如下。

```java
private void initReaderAddUi() {

    for (int i = 0; i < Constant.READER_YTPES.length; i++) {
        cb_rd_add_reader_type.getItems().addAll(Constant.READER_YTPES[i]);
        cb_rd_alter_reader_type.getItems().addAll(Constant.READER_YTPES[i]);
        cb_rd_delete_reader_type.getItems().addAll(Constant.READER_YTPES[i]);
    }
    cb_rd_add_reader_type.getSelectionModel().selectFirst();
    cb_rd_alter_reader_type.getSelectionModel().selectFirst();
    cb_rd_delete_reader_type.getSelectionModel().selectFirst();

    for (int i = 0; i < Constant.SEX.length; i++) {
        cb_rd_add_reader_sex.getItems().addAll(Constant.SEX[i]);
        cb_rd_alter_reader_sex.getItems().addAll(Constant.SEX[i]);
        cb_rd_delete_reader_sex.getItems().addAll(Constant.SEX[i]);
    }
    cb_rd_add_reader_sex.getSelectionModel().selectFirst();
    cb_rd_alter_reader_sex.getSelectionModel().selectFirst();
    cb_rd_delete_reader_sex.getSelectionModel().selectFirst();

    RequiredFieldValidator validator_ts_book_add = new RequiredFieldValidator();
```

```
    validator_ts_book_add.setMessage("请输入...");
    tf_rd_alter_reader_search_id.getValidators().add(validator_ts_book_add);
    tf_rd_alter_reader_search_id.focusedProperty().addListener((o,oldVal,newVal)->{
        if(!newVal) tf_rd_alter_reader_search_id.validate();
    });

    tf_rd_delete_reader_search_id.getValidators().add(validator_ts_book_add);
    tf_rd_delete_reader_search_id.focusedProperty().addListener((o,oldVal,newVal)->{
        if(!newVal) tf_rd_delete_reader_search_id.validate();
    });

}
```

2)添加新的读者

通过函数 add_new_reader()添加新的读者,具体实现代码如下。

```
@FXML
public void add_new_reader() {
    if (!tf_rd_add_reader_id.getText().equals("") && !tf_rd_add_reader_name.getText().equals("") && !tf_rd_add_reader_numbers.getText().equals("")
            && !tf_rd_add_reader_days.getText().equals("") &&
            !cb_rd_add_reader_type.getSelectionModel().getSelectedItem().toString().equals("") && !cb_rd_add_reader_sex.getSelectionModel().getSelectedItem().toString().equals("")) {
        Reader reader = new Reader();
        reader.setId(tf_rd_add_reader_id.getText());
        reader.setName(tf_rd_add_reader_name.getText());
        reader.setPassword("123456");//默认密码
        reader.setType(cb_rd_add_reader_type.getSelectionModel().
                    getSelectedItem().toString());
        reader.setSex(cb_rd_add_reader_sex.getSelectionModel().
                    getSelectedItem().toString());
        reader.setMax_num(Integer.parseInt(tf_rd_add_reader_numbers.getText()));
        reader.setDays_num(Integer.parseInt(tf_rd_add_reader_days.getText()));
        reader.setForfeit(0);

        Boolean isok = DataBaseUtil.addNewReader(reader);
        if (isok) {
            System.out.println("add ok");
            Alert alert = new Alert(Alert.AlertType.CONFIRMATION);
            alert.setAlertType(Alert.AlertType.INFORMATION);
            alert.setContentText("添加成功! ");
            alert.setTitle("添加成功! ");
            alert.show();
            rd_reader_add_clear();
        } else {
            Alert alert = new Alert(Alert.AlertType.CONFIRMATION);
            alert.setAlertType(Alert.AlertType.ERROR);
```

```
                alert.setContentText("添加失败! ");
                alert.setTitle("添加失败! ");
                alert.show();
            }
        } else {
            Alert alert = new Alert(Alert.AlertType.CONFIRMATION);
            alert.setAlertType(Alert.AlertType.ERROR);
            alert.setContentText("信息不完整! ");
            alert.setTitle("添加错误! ");
            alert.show();
        }
    }
```

添加读者信息界面的执行效果如图 3-15 所示。

图 3-15　添加读者信息界面

3) 修改读者信息

通过函数 alter_rd_reader() 修改读者的基本信息，具体实现代码如下。

```
    @FXML
    public void alter_rd_reader() {
        if (!tf_rd_alter_reader_id.getText().equals("")
&& !tf_rd_alter_reader_name.getText().equals("")
&& !tf_rd_alter_reader_numbers.getText().equals("")
&& !tf_rd_alter_reader_days.getText().equals("") &&
            !cb_rd_alter_reader_type.getSelectionModel().getSelectedItem().
                toString().equals("") && !cb_rd_alter_reader_sex.getSelectionModel().
                getSelectedItem().toString().equals("")) {
            Reader reader = new Reader();
            reader.setId(tf_rd_alter_reader_id.getText());
            reader.setName(tf_rd_alter_reader_name.getText());
            if (tgBtn_rd_alter_reader_password_reset.isPressed()) {
```

```java
                reader.setPassword("123456");//默认密码
            } else {
                reader.setPassword(rd_reader_alter_password);//原密码
            }
            reader.setType(cb_rd_alter_reader_type.getSelectionModel().
                    getSelectedItem().toString());
            reader.setSex(cb_rd_alter_reader_sex.getSelectionModel().
                    getSelectedItem().toString());
            reader.setMax_num(Integer.parseInt(tf_rd_alter_reader_numbers.getText()));
            reader.setDays_num(Integer.parseInt(tf_rd_alter_reader_days.getText()));
            reader.setForfeit(0);
            
            Boolean isok = DataBaseUtil.alterReader(reader);
            if (isok) {
                System.out.println("add ok");
                Alert alert = new Alert(Alert.AlertType.CONFIRMATION);
                alert.setAlertType(Alert.AlertType.INFORMATION);
                alert.setContentText("修改成功! ");
                alert.setTitle("修改成功! ");
                alert.show();
                rd_reader_alter_clear();
            } else {
                Alert alert = new Alert(Alert.AlertType.CONFIRMATION);
                alert.setAlertType(Alert.AlertType.ERROR);
                alert.setContentText("修改失败! ");
                alert.setTitle("修改失败! ");
                alert.show();
            }
    } else {
        Alert alert = new Alert(Alert.AlertType.CONFIRMATION);
        alert.setAlertType(Alert.AlertType.ERROR);
        alert.setContentText("信息不完整! ");
        alert.setTitle("修改错误! ");
        alert.show();
    }
}

private String rd_reader_alter_password = "123456";
```

通过函数rd_reader_alter_search()查询某名读者的信息，输入编号后，可以显示这名读者的详细信息。具体实现代码如下。

```java
@FXML
public void rd_reader_alter_search() {
    if (!tf_rd_alter_reader_search_id.getText().equals("")) {
        Reader reader = DataBaseUtil.getReader(tf_rd_alter_reader_search_
                id.getText().trim());
```

```
            if (reader != null) {
                tf_rd_alter_reader_id.setText(reader.getId());
                tf_rd_alter_reader_name.setText(reader.getName());
                if (reader.getType().equals("教师")) {
                    cb_rd_alter_reader_type.getSelectionModel().selectFirst();
                } else if (reader.getType().equals("学生")) {
                    cb_rd_alter_reader_type.getSelectionModel().select(1);
                } else {
                    cb_rd_alter_reader_type.getSelectionModel().select(2);
                }
                if (reader.getSex().equals("男")) {
                    cb_rd_alter_reader_sex.getSelectionModel().selectFirst();
                } else {
                    cb_rd_alter_reader_sex.getSelectionModel().select(1);
                }
                rd_reader_alter_password = reader.getPassword();
                tf_rd_alter_reader_numbers.setText(reader.getMax_num()+"");
                tf_rd_alter_reader_days.setText(reader.getDays_num()+"");
            } else {
                tf_rd_alter_reader_search_id.setText("");
                tf_rd_alter_reader_search_id.validate();
            }
        }
    }
```

修改读者信息界面的执行效果如图 3-16 所示。

图 3-16　修改读者信息界面

4) 删除读者信息

首先通过函数 rd_reader_delete_search()查询某条要删除的读者信息，具体实现代码如下。

```java
@FXML
public void rd_reader_delete_search() {
    if (!tf_rd_delete_reader_search_id.getText().equals("")) {
        Reader reader = DataBaseUtil.getReader(tf_rd_delete_reader_search_
                        id.getText().trim());
        if (reader != null) {
            tf_rd_delete_reader_id.setText(reader.getId());
            tf_rd_delete_reader_name.setText(reader.getName());
            if (reader.getType().equals("教师")) {
                cb_rd_delete_reader_type.getSelectionModel().selectFirst();
            } else if (reader.getType().equals("学生")) {
                cb_rd_delete_reader_type.getSelectionModel().select(1);
            } else {
                cb_rd_delete_reader_type.getSelectionModel().select(2);
            }
            if (reader.getSex().equals("男")) {
                cb_rd_delete_reader_sex.getSelectionModel().selectFirst();
            } else {
                cb_rd_delete_reader_sex.getSelectionModel().select(1);
            }
            tf_rd_delete_reader_numbers.setText(reader.getMax_num()+"");
            tf_rd_delete_reader_days.setText(reader.getDays_num()+"");
        } else {
            tf_rd_delete_reader_search_id.setText("");
            tf_rd_delete_reader_search_id.validate();
        }
    }
}
```

然后通过函数 delete_rd_reader()删除这条被查询到的读者信息，具体实现代码如下。

```java
@FXML
public void delete_rd_reader() {
    if (!tf_rd_delete_reader_id.getText().equals("")) {

        Alert alert = new Alert(Alert.AlertType.CONFIRMATION);
        alert.setAlertType(Alert.AlertType.CONFIRMATION);
        alert.setContentText("确认删除？");
        alert.setTitle("确认删除！");
        alert.showAndWait();
        ButtonType type = alert.getResult();
        System.out.println("type="+type.getText());
        if (type == ButtonType.OK) {
            Boolean isok = DataBaseUtil.deleteReader(tf_rd_delete_reader_id.getText());
```

```java
            if (isok) {
                System.out.println("add ok");
                Alert alert1 = new Alert(Alert.AlertType.CONFIRMATION);
                alert1.setAlertType(Alert.AlertType.INFORMATION);
                alert1.setContentText("删除成功！");
                alert1.setTitle("删除成功！");
                alert1.show();
                rd_reader_delete_clear();
            } else {
                Alert alert2 = new Alert(Alert.AlertType.CONFIRMATION);
                alert2.setAlertType(Alert.AlertType.ERROR);
                alert2.setContentText("删除失败！");
                alert2.setTitle("删除失败！");
                alert2.show();
            }
        }
    } else {
        Alert alert = new Alert(Alert.AlertType.CONFIRMATION);
        alert.setAlertType(Alert.AlertType.ERROR);
        alert.setContentText("信息不完整！");
        alert.setTitle("删除错误！");
        alert.show();
    }
}
```

删除读者信息界面的执行效果如图 3-17 所示。

图 3-17　删除读者信息界面

5) 显示全部读者信息

通过函数 getAllReaders() 获取并显示系统内的全部读者信息，具体实现代码如下。

```java
@FXML
public void getAllReaders() {
```

```
            ObservableList<Reader> readers = DataBaseUtil.getAllReaders();
            if (readers != null) {
                tbv_reader.setItems(readers);
            }else {
//              tbv_reader.setAccessibleText("无记录");
            }
```

显示全部读者信息界面的执行效果如图 3-18 所示。

图 3-18 显示全部读者信息界面

3.8.2 图书信息管理

1) 添加新类别

本系统的图书信息管理包含图书类别管理、添加图书、修改图书和删除图书等功能，其中图书类别管理只有一个添加新类别功能，具体实现代码如下。

```
        private void updateBookType() {
            Constant.BOOK_TYPE = DataBaseUtil.getBookType();
            
            int size = cb_ts_book_type.getItems().size();
            for (int i = 0; i < size; i++) {
                cb_ts_book_type.getItems().remove(0);
                cb_ts_add_book_type.getItems().remove(0);
                cb_ts_alter_book_type.getItems().remove(0);
            }
            
            Set set = Constant.BOOK_TYPE.keySet();
            Iterator iter = set.iterator();
            while (iter.hasNext()) {
```

```java
            String key = (String) iter.next();
            cb_ts_book_type.getItems().addAll(key);
            cb_ts_add_book_type.getItems().addAll(key);
            cb_ts_alter_book_type.getItems().addAll(key);
            cb_ts_delete_book_type.getItems().addAll(key);
        }
        //选择第一个
        cb_ts_book_type.getSelectionModel().selectFirst();
        cb_ts_add_book_type.getSelectionModel().selectFirst();
        cb_ts_alter_book_type.getSelectionModel().selectFirst();
    }

    public int getBookTypeSelectNumber(int id) {
        int number = 0;
        Set set = Constant.BOOK_TYPE.keySet();
        Iterator iter = set.iterator();
        while (iter.hasNext()) {
            String key = (String) iter.next();
            if (id == Constant.BOOK_TYPE.get(key)) {
                return number;
            }
            number++;
        }
        return 0;
    }

    public String getBookTypeAccordingId(int id) {
        Set set = Constant.BOOK_TYPE.keySet();
        Iterator iter = set.iterator();
        while (iter.hasNext()) {
            String key = (String) iter.next();
            if (id == Constant.BOOK_TYPE.get(key)) {
                return key;
            }
        }
        return "";
    }

    private int getBookIdAccordingToSelectNumber(int selectedIndex) {
        int number = 0;
        Set set = Constant.BOOK_TYPE.keySet();
        Iterator iter = set.iterator();
        while (iter.hasNext()) {
            String key = (String) iter.next();
            if (selectedIndex == number) {
                return Constant.BOOK_TYPE.get(key);
            }
            number++;
```

```
        }
        return 0;
    }
```

图书类别信息管理界面的执行效果如图3-19所示。

图3-19　图书类别信息管理界面

2）添加图书信息

通过函数ts_book_add()实现添加图书信息功能，具体实现代码如下。

```
    public void ts_book_add() {
        System.out.println("info====>  "+tf_ts_add_book_id.getText() + tf_ts_add_
book_name.getText()+tf_ts_add_book_author.getText()+tf_ts_add_book_translator.
getText()+tf_ts_add_book_publisher.getText()+tf_ts_add_book_price.getText()
            +tf_ts_add_book_stock.getText()+cb_ts_add_book_type.getSelectionModel().
getSelectedItem().toString()+dp_ts_add_book_publish_time.getEditor().getText());
        if (!tf_ts_add_book_id.getText().equals("")
&& !tf_ts_add_book_name.getText().equals("")
&& !tf_ts_add_book_author.getText().equals("")
&& !tf_ts_add_book_translator.getText().equals("")
&& !tf_ts_add_book_publisher.getText().equals("") &&
            !tf_ts_add_book_price.getText().equals("") && !tf_ts_add_book_
stock.getText().equals("") && !cb_ts_add_book_type.getSelectionModel().
getSelectedItem().toString().equals("") && !dp_ts_add_book_publish_
time.getEditor().getText().equals("")) {
            Book book = new Book();
            book.setId(tf_ts_add_book_id.getText());
            book.setName(tf_ts_add_book_name.getText());
            book.setType(Constant.BOOK_TYPE.get(cb_ts_add_book_
            type.getSelectionModel().getSelectedItem().toString()));
            book.setAuthor(tf_ts_add_book_author.getText());
            book.setTranslator(tf_ts_add_book_translator.getText());
            book.setPublisher(tf_ts_add_book_publisher.getText());
            book.setPublishTime(dp_ts_add_book_publish_time.getEditor().getText());
            book.setStock(Integer.parseInt(tf_ts_add_book_stock.getText()));
```

```
            book.setPrice(Double.parseDouble(tf_ts_add_book_price.getText()));
            Boolean isok = DataBaseUtil.addNewBook(book);
            if (isok) {
                System.out.println("add ok");
                Alert alert = new Alert(Alert.AlertType.CONFIRMATION);
                alert.setAlertType(Alert.AlertType.INFORMATION);
                alert.setContentText("添加成功!");
                alert.setTitle("添加成功!");
                alert.show();
                ts_book_add_clear();
            } else {
                Alert alert = new Alert(Alert.AlertType.CONFIRMATION);
                alert.setAlertType(Alert.AlertType.ERROR);
                alert.setContentText("添加失败!");
                alert.setTitle("添加失败!");
                alert.show();
            }
        } else {
            Alert alert = new Alert(Alert.AlertType.CONFIRMATION);
            alert.setAlertType(Alert.AlertType.ERROR);
            alert.setContentText("信息不完整!");
            alert.setTitle("添加错误!");
            alert.show();
        }
    }
```

添加图书信息界面的执行效果如图 3-20 所示。

图 3-20　添加图书信息界面

修改图书和删除图书功能跟前面讲解的读者修改和读者删除功能类似，具体实现方法也类似，在此不再赘述。

3.8.3 借书处理模块

（1）本项目的核心是借书和还书模块，我们首先看借书模块的实现过程。通过函数 js_confirm_start() 实现单击 "确认" 按钮后的借书功能，具体实现代码如下。

```
@FXML
public void js_confirm_start() {
    if (!tf_js_book_name.getText().equals("")
&& !tf_js_reader_name.getText().equals("")) {
        boolean isBorrow = DataBaseUtil.addNewBorrow(tf_js_book_id.getText(),
tf_js_reader_id.getText(), lb_js_reader_jieshu_date.getText(),
lb_js_reader_huanshu_date.getText(), 0);
        if (isBorrow) {
            Alert alert = new Alert(Alert.AlertType.CONFIRMATION);
            alert.setAlertType(Alert.AlertType.INFORMATION);
            alert.setContentText("借书成功！");
            alert.setTitle("借书成功！");
            alert.show();
            tf_js_book_id.setText("");
            tf_js_reader_id.setText("");
            clear_js_book();
            clear_js_reader();
        }
    } else {
        tf_js_book_id.setText("");
        tf_js_reader_id.setText("");
        clear_js_book();
        clear_js_reader();
        tf_js_book_id.validate();
        tf_js_reader_id.validate();
    }
}
```

（2）编写函数 tf_js_book_id_keyEvent()，输入图书编号并按回车键，会显示对应的图书信息，具体实现代码如下。

```
public void tf_js_book_id_keyEvent(KeyEvent keyEvent) {
    if (keyEvent.getCode().equals(KeyCode.ENTER)) {
        Book book = DataBaseUtil.getBook(tf_js_book_id.getText());
        if (book != null) {
//            tf_js_book_id.setText(book.getId());
            tf_js_book_name.setText(book.getName());
            tf_js_book_publisher.setText(book.getPublisher());
            tf_js_book_publish_time.setText(book.getPublishTime().toString());
        } else {
```

```
                clear_js_book();
            }
        }
    }
}
```

(3) 编写函数 js_reader_id_keyEvent()，输入读者编号后显示对应的读者信息，具体实现代码如下。

```
public void tf_js_reader_id_keyEvent(KeyEvent keyEvent) {
    if (keyEvent.getCode().equals(KeyCode.ENTER)) {
        Reader reader = DataBaseUtil.getReader(tf_js_reader_id.getText());
        if (reader != null) {
//            tf_js_reader_id.setText(reader.getId());
            tf_js_reader_name.setText(reader.getName());
            tf_js_reader_type.setText(reader.getType());
            tf_js_reader_sex.setText(reader.getSex());
            lb_js_reader_huanshu_date.setText(DateUtils.getAfterDay
(lb_js_reader_jieshu_date.getText(), reader.getDays_num()));
        } else {
            clear_js_reader();
        }
    }
}
```

借书界面的执行效果如图 3-21 所示。

图 3-21　借书界面

3.8.4　还书处理模块

(1) 通过函数 tf_hs_reader_id_keyEvent()监听读者的 ID 号输入，输入 ID 号并按回车键后会显示对应的还书信息，具体实现代码如下。

```java
private void tf_hs_reader_id_keyEvent(KeyEvent keyEvent) {
    if (keyEvent.getCode().equals(KeyCode.ENTER)) {
        Reader reader = DataBaseUtil.getReader(tf_hs_reader_id.getText());
        //如果不为空，则进行
        if (reader != null) {
            tf_hs_reader_name.setText(reader.getName());
            tf_hs_reader_type.setText(reader.getType());
            tf_hs_reader_sex.setText(reader.getSex());
            getReaderBorrowRecord(reader.getId());
        } else {
            clear_hs_reader();
        }
    }
}
```

（2）编写函数 tf_hs_reader_id_keyEvent()，用于在还书成功后刷新界面，具体实现代码如下。

```java
private void tf_hs_reader_id_keyEvent() {
    Reader reader = DataBaseUtil.getReader(tf_hs_reader_id.getText());
    //如果不为空，则进行
    if (reader != null) {
        tf_hs_reader_name.setText(reader.getName());
        tf_hs_reader_type.setText(reader.getType());
        tf_hs_reader_sex.setText(reader.getSex());
        getReaderBorrowRecord(reader.getId());
    } else {
        clear_hs_reader();
    }
}
```

（3）编写函数 getReaderBorrowRecord()，用于获取读者的借书信息，具体实现代码如下。

```java
public void getReaderBorrowRecord(String id) {
    ObservableList<borrow_record> borrows = DataBaseUtil.getBorrowRecord(id);
    if (borrows != null) {
        tbv_huanshu_record.setItems(borrows);
    }else {

    }
}
```

（4）编写函数 huanshu_start()，用于单击"确认"按钮后实现还书操作，具体实现代码如下。

```java
@FXML
private void huanshu_start() {
```

```
            if (!tf_hs_reader_name.getText().trim().equals("")) {
                if (!tf_hs_book_id.getText().trim().equals("")) {
                    double isok = DataBaseUtil.backBook(tf_hs_reader_id.getText().trim(),
                    tf_hs_book_id.getText().trim());
                    if (isok != -1) {
                        tf_hs_reader_id_keyEvent();
                        tf_hs_book_id.setText("");
                        tf_hs_book_name.setText("");
                        System.out.println("huanshu ok");
                        Alert alert = new Alert(Alert.AlertType.CONFIRMATION);
                        alert.setAlertType(Alert.AlertType.INFORMATION);
                        alert.setContentText("还书成功！  超期罚款为： " + isok);
                        alert.setTitle("还书成功！ ");
                        alert.show();
                    } else {
                        Alert alert = new Alert(Alert.AlertType.CONFIRMATION);
                        alert.setAlertType(Alert.AlertType.ERROR);
                        alert.setContentText("还书失败！");
                        alert.setTitle("还书失败！ ");
                        alert.show();
                    }
                } else {
                    tf_hs_book_id.validate();
                }
            } else {
                tf_hs_reader_id.validate();
            }
        }
```

还书界面的执行效果如图 3-22 所示。

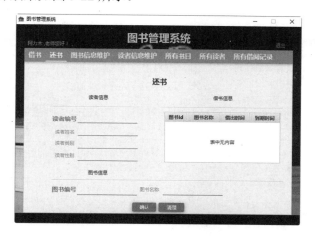

图 3-22　还书界面

3.9 数据操作

本书前面介绍的功能模块都涉及了和数据库操作相关的功能，例如添加、修改、删除图书，借书和还书处理等。为了实现业务逻辑和数据操作的分离，我们将数据库数据的具体操作封装到了一个单独文件 DataBaseUtil.java 中。本节将简要讲解这个文件的具体实现流程。

扫码看视频

3.9.1 用户登录验证

（1）编写函数 checkReader()，功能是建立和系统数据库的连接，查询当前登录用户是不是系统合法用户，合法就登录成功，不合法则提示非法登录信息。函数 checkReader() 的具体实现代码如下。

```java
/**
 * 读者登录检查
 * @param account
 * @param password
 * @return
 */
public static boolean checkReader(String account, String password) {
    boolean checkbool = false;
    try {
        Connection con = null; //定义一个MySQL连接对象
        con = DriverManager.getConnection("jdbc:mysql://localhost/tushuguanli_
            database_test?" +"user=root&password=66688888&useSSL=
            false&serverTimezone=GMT");
        //连接本地MySQL
        Statement stmt; //创建声明
        stmt = con.createStatement();

        String password_fromDb;
        String selectSql = "select pass from Reader where Id='"+account+"'";
        ResultSet selectRes = stmt.executeQuery(selectSql);
        if (selectRes.next()) {
            password_fromDb = selectRes.getString("pass");
            if (password_fromDb.equals(password)) {
                checkbool = true;
            }
        }
        con.close();
    } catch (Exception e) {
```

```
            System.out.print("读者登录检查----checkReader----MySQL ERROR:" + e.getMessage());
        }
        return checkbool;
    }
```

(2) 编写函数 checkUser()，功能是建立和系统数据库的连接，查询当前管理员是否已经登录系统。具体代码如下所示。

```
public static boolean checkUser(String account, String password) {
    boolean checkbool = false;
    try {
        Connection con = null; //定义一个MySQL连接对象
        con =DriverManager.getConnection("jdbc:mysql://localhost/tushuguanli_
            database_test?" +"user=root&password=66688888&useSSL=
            false&serverTimezone=GMT");
        //连接本地MySQL
        Statement stmt; //创建声明
        stmt = con.createStatement();

        String password_fromDb;
        String selectSql = "SELECT pass FROM user where Id='"+account+"'";
        ResultSet selectRes = stmt.executeQuery(selectSql);
        if (selectRes.next()) {
            password_fromDb = selectRes.getString("pass");
            if (password_fromDb.equals(password)) {
                checkbool = true;
            }
        }
        con.close();
    } catch (Exception e) {
        System.out.print("管理员登录检查----checkUser----MySQL ERROR:" + e.getMessage());
    }
    return checkbool;
}
```

3.9.2 获取图书信息

编写函数 getBook()，功能是查询系统数据库中的图书信息，并将图书信息显示在窗体界面中。函数 getBook() 的具体实现代码如下。

```
public static Book getBook(String id) {
    Book book = null;
    try {
        Connection con = null; //定义一个MySQL连接对象
        con =DriverManager.getConnection
                ("jdbc:mysql://localhost/tushuguanli_database_test?" +
```

```
                "user=root&password=66688888&useSSL=false&serverTimezone=GMT");
        //连接本地MySQL
        Statement stmt; //创建声明
        stmt = con.createStatement();

        String selectSql = "SELECT * FROM book where Id='"+id+"'";
        ResultSet selectRes = stmt.executeQuery(selectSql);
        if (selectRes.next()) {
            book = new Book();
            book.setId(selectRes.getString("Id"));
            book.setName(selectRes.getString("name"));
            book.setType(selectRes.getInt("type"));
            book.setAuthor(selectRes.getString("author"));
            book.setTranslator(selectRes.getString("translator"));
            book.setPublisher(selectRes.getString("publisher"));
            book.setPublishTime(selectRes.getDate("publish_time").toString());
            book.setStock(selectRes.getInt("stock"));
            book.setPrice(selectRes.getDouble("price"));
        }
        con.close();
        return book;
    } catch (Exception e) {
        System.out.print("book 获取检查----getBook----MySQL ERROR:" + e.getMessage());
    }
    return book;
}
```

3.9.3 获取读者信息

编写函数 getReader()，功能是根据编号查询数据库中该名读者的信息。函数 getReader() 的具体实现代码如下。

```
public static Reader getReader(String id) {
    Reader reader = null;
    try {
        Connection con = null; //定义一个MySQL连接对象
        con =DriverManager.getConnection("jdbc:mysql://localhost/
            tushuguanli_database_test?" +
            "user=root&password=66688888&useSSL=false&serverTimezone=GMT");
        //连接本地MySQL
        Statement stmt; //创建声明
        stmt = con.createStatement();

        String selectSql = "SELECT * FROM reader where Id='"+id+"'";
        ResultSet selectRes = stmt.executeQuery(selectSql);
        if (selectRes.next()) {
```

```
            reader = new Reader();
            reader.setId(selectRes.getString("Id"));
            reader.setName(selectRes.getString("name"));
            reader.setPassword(selectRes.getString("pass"));
            reader.setType(selectRes.getString("type"));
            reader.setSex(selectRes.getString("sex"));
            reader.setMax_num(selectRes.getInt("max_num"));
            reader.setDays_num(selectRes.getInt("days_num"));
            reader.setForfeit(selectRes.getDouble("forfeit"));
        }
        con.close();
        return reader;
    } catch (Exception e) {
        System.out.print("reader 获取检查----getReader----MySQL ERROR:" +
                    e.getMessage());
    }
    return reader;
}
```

3.9.4 添加借阅记录信息

编写函数 addNewBorrow()，功能是向数据库中添加新的图书借阅信息。在添加借阅记录时，需要使用函数 isAlreadyBorrowed() 验证当前图书是否已经被借阅。其中函数 addNewBorrow() 的具体实现代码如下。

```
    public static boolean addNewBorrow(String bookId, String readerId, String
jieshu_date, String huanshu_date, int isBack) {
        try {
            Connection con = null; //定义一个MySQL 连接对象
            con =DriverManager.getConnection("jdbc:mysql://localhost/tushuguanli_
                database_test?" +
                    "user=root&password=66688888&useSSL=false&serverTimezone=GMT");
            //连接本地MySQL
            Statement stmt; //创建声明
            stmt = con.createStatement();
            //检查书库存
            Book book = getBook(bookId);
            boolean isAlreadyBorrowed = isAlreadyBorrowed(readerId,bookId);
            if (book.getStock() <= 0 || isAlreadyBorrowed) {
                return false;
            }
            //继续借书
            String updateSql = "insert into borrow (book_id,reader_id,borrow_date,
                    back_date,is_back) values ('"+bookId+"','"+readerId+"',
                    '"+jieshu_date+"', '"+huanshu_date+"','"+isBack+"')";
```

```
            int selectRes = stmt.executeUpdate(updateSql);
            if (selectRes != 0) {
                return true;
            }
            con.close();
        } catch (Exception e) {
            System.out.print("addBorrow检查----addNewBorrow----MySQL ERROR:" +
                            e.getMessage());
        }
        return false;
    }
```

函数 isAlreadyBorrowed()的具体实现代码如下。

```
    private static boolean isAlreadyBorrowed(String readerId, String bookId) {
        try {
            Connection con = null; //定义一个MySQL连接对象
            con =DriverManager.getConnection("jdbc:mysql://localhost/tushuguanli_
                database_test?" +
                    "user=root&password=66688888&useSSL=false&serverTimezone=GMT");
            //连接本地MySQL
            Statement stmt; //创建声明
            stmt = con.createStatement();
            //继续借书
            String selectSql = "select * from borrow where reader_id='"+readerId+"'
                            and book_id='"+bookId+"'";

            ResultSet selectRes = null;
            selectRes = stmt.executeQuery(selectSql);
            if (selectRes.next()) {
                return true;
            }
            con.close();
        } catch (Exception e) {
            System.out.print("isAlreadyBorrowed----isAlreadyBorrowed----MySQL
                            ERROR:" + e.getMessage());
        }
        return false;
    }
```

3.9.5 添加新书信息

编写函数 addNewBook()，功能是向数据库中添加新的图书信息。函数 addNewBook()的具体实现代码如下。

```java
    public static boolean addNewBook(Book book) {
        try {
            Connection con = null; //定义一个MySQL连接对象
            con =DriverManager.getConnection("jdbc:mysql://localhost/tushuguanli_
                database_test?" +
                "user=root&password=66688888&useSSL=false&serverTimezone=GMT");
            //连接本地MySQL
            Statement stmt; //创建声明
            stmt = con.createStatement();

            String updateSql = "insert into book
(Id,name,type,author,translator,publisher,publish_time,stock,price) values ('" +
book.getId() + "','" + book.getName() + "','" + book.getType() + "','" +
book.getAuthor() + "','" +book.getTranslator()+ "','" +book.getPublisher()+ "','"
+ book.getPublishTime()+ "','" + book.getStock()+ "','" + book.getPrice() + "')";
            int selectRes = stmt.executeUpdate(updateSql);
            if (selectRes != 0) {
                return true;
            }
            con.close();
        } catch (Exception e) {
            System.out.print("addBook检查----addNewBook----MySQL ERROR:" +
                            e.getMessage());
        }
        return false;

    }
```

除此之外，在文件 DataBaseUtil.java 中还提供了其他功能模块的数据库处理功能，例如查询用户信息、修改图书信息、删除图书信息、添加读者信息、修改读者信息，等等。为了节省篇幅，书中不再讲解这些功能的具体实现过程。

第 4 章 物业管理系统

随着世界经济的快速发展，房地产业在经济发展中扮演着重要的角色，各城市规模不断扩大，城市人口日益增多，怎么管理好社区成为当今居民热议的话题。实现社区信息化管理，是宜居城市的必然趋势。使用物业管理系统管理不同种类且数量繁多的小区，可以提高物业管理工作的效率，减少工作中可能出现的错误，为居民提供更好的服务，是提高城市生活水平的重要组成部分。本章将介绍使用 Java 语言开发物业管理系统的方法，旨在让读者牢固掌握 SQL 后台数据库的建立、维护以及前台应用程序的开发等方法。本章项目由 IntelliJ IDEA+Swing+AWT+MySQL 实现。

4.1 背景介绍

随着市场经济的发展和人们生活水平的提高，住宅小区已经成为人们安家置业的首选，几十万到几百万人口的小区住宅比比皆是。人们不但对住宅本身的美观、质量等要求越来越高，同时对小区物业的服务和管理要求也越来越高。这就要求小区管理者要对物业进行宏观的和微观的管理，其中最好的办法是用计算机操作小区物业管理系统，使管理者和业主对小区中的事务得到更方便、更快捷、更满意的答复。

扫码看视频

4.2 系统分析和设计

本项目的系统规划和设计主要完成以下工作。
- 系统需求分析。
- 设计流程分析。
- 系统运行流程。

扫码看视频

4.2.1 系统需求分析

物业管理系统的用户主要是物业公司的办公人员，需要具备的基本功能如下。

1) 基础信息管理功能模块

基础信息管理模块的主要功能包括收费单价录入、房屋信息录入、楼宇信息录入和小区信息录入四个部分。

2) 费用报表模块

针对物业公司的需求，主要费用报表信息包括物业费用报表、业主费用报表、气费报表查询、电费报表查询和水费报表查询。

3) 消费指数模块

消费指数模块的功能是录入各种指数信息，包括物业消费指数和业主消费指数两个部分。

根据上述需求分析，设计系统的体系结构，如图 4-1 所示。在这个体系结构中，每一个叶节点是一个最小的功能模块，每一个功能模块都需要针对不同的表完成相应的数据库操作，主要包括添加记录、删除记录、查询记录和更新记录等。

4.2.2 设计流程分析

本项目的具体设计流程如图 4-2 所示。

第 4 章　物业管理系统

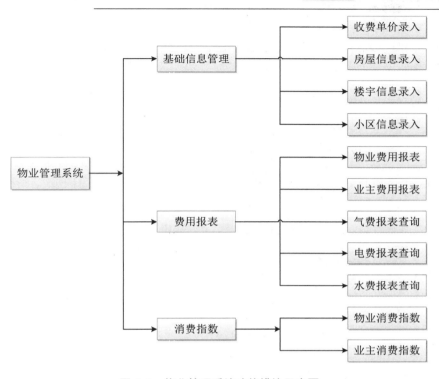

图 4-1　物业管理系统功能模块示意图

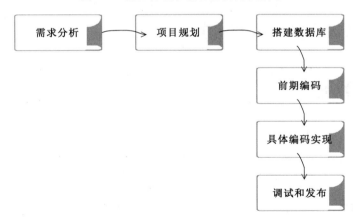

图 4-2　设计流程图

整个项目的具体操作流程是：项目规划→数据库设计→框架设计→系统基础信息管理、用户管理、消费指数和各项费用管理。

本项目包括后台数据库的建立以及前端应用程序的开发两个方面。其中后台数据库采用 MySQL，同时在程序设计中应用面向对象思想，这也是本项目的优势和特色。本项目的具体实现流程如图 4-3 所示。

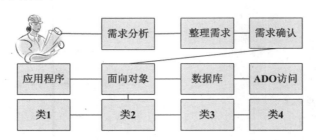

图 4-3　具体实现流程

4.2.3　系统模拟流程

模拟系统的运行流程是：运行系统后，首先会弹出用户登录对话框，对用户的身份进行认证并确定用户的合法性。如果使用本系统 admin 用户(系统管理员)登录，可以在系统管理菜单下对系统数据进行添加、修改和删除等操作。

在系统初始化时，系统管理员的默认用户名为 admin，密码为 admin。登录系统后，首先需要添加基础信息。基础信息包括小区信息、楼宇信息、房屋信息和收费项目单价。基础信息是物业管理系统的基础数据，它为物业管理系统的其他模块提供数据参考。其中，小区信息包括小区名、小区编号、地址等，楼宇信息包括小区编号、楼宇编号、楼层数、面积属性描述等，房屋信息包括业主姓名、电话、地址和性别等。

4.3　数据库设计

本项目的开发工作主要包括后台数据库的建立、测试数据的录入以及前台应用程序的开发三个方面。数据库设计是系统开发的一个重要组成部分，数据库设计得好坏会直接影响程序编码的复杂程度。

4.3.1　选择数据库

在开发物业管理系统时，需要根据用户需求、系统功能和性能要求等因素来设计和优化数据。后台数据库需要考虑用户需求、系统功能和性能要求等因素。数据库是数据管理的最新技术，是计算机科学的重要分支。由于数据库具有数据结构化、最低冗余度、较高的程序与数据独立性，以及易于扩充、易于编制应用程序等优点，许多管理系统都是建立在数据库基础上的。考虑到实际情况，同时 MySQL 是一款免费的数据库软件，所以本项目将使用 MySQL 作为后台数据库管理平台。

数据库访问技术决定了整个项目的访问效率，本项目将使用 ADO 数据库访问技术。

4.3.2 数据库结构设计

具体的数据库设计需要参考需求分析文档。由需求分析的规划可知，整个项目对象有 7 种信息，所以对应的数据库也需要包含这 7 种信息，因此本项目需要包含以下 7 个数据库表。

- user：用户信息表。
- price_type：费用类型表。
- master_info：业主信息表。
- master_use：业主消费指数表。
- community_info：小区信息表。
- building_info：楼宇信息表。
- community_use：物业消费指数表。

(1) 用户信息表(user)用来保存用户信息，其结构如表 4-1 所示。

表 4-1 用户信息表结构

编 号	字段名称	数据类型	说 明
1	uname	varchar12	用户名
2	paswrd	varchar20	密码
3	purview	smallint6	权限

(2) 费用类型表(price_type)用来保存费用类型信息，其结构如表 4-2 所示。

表 4-2 费用类型表结构

编 号	字段名称	数据结构	说 明
1	charge_id	smallint6	费用 ID
2	charge_name	varchar20	费用名称
3	unit_price	double	价格

(3) 业主信息表(master_info)用来保存系统内的业主信息，其结构如表 4-3 所示。

(4) 业主消费指数表(master_use)用来保存业主消费指数信息，其结构如表 4-4 所示。

(5) 小区信息表(community_info)用来保存小区信息，其结构如表 4-5 所示。

表 4-3 业主信息表结构

编 号	字段名称	数据结构	说 明
1	district_id	smallint6	小区 ID
2	building_id	smallint6	楼宇 ID
3	room_id	smallint6	业主 ID
4	area	double	面积
5	status	varchar9	状态
6	purpose	varchar10	用途
7	oname	varchar20	业主名称
8	sex	varchar7	性别
9	id_num	varchar18	证件号
10	address	varchar60	地址
11	phone	double	电话

表 4-4 业主消费指数表结构

编 号	字段名称	数据结构	说 明
1	district_id	smallint6	小区 ID
2	building_id	smallint6	楼宇 ID
3	room_id	smallint6	业主 ID
4	date	int	日期
5	water_reading	double	用水量
6	elec_reading	double	用电量
7	gas_reading	double	用气量

表 4-5 小区信息表结构

编 号	字段名称	数据结构	说 明
1	district_id	smallint6	小区 ID
2	district_name	varchar12	小区名称
3	address	varchar60	地址
4	floor_space	double	小区面积

(6) 楼宇信息表(building_info)用来保存楼宇信息，其结构如表 4-6 所示。

表 4-6　楼宇信息表结构

编 号	字段名称	数据结构	说 明
1	district_id	smallint6	小区 ID
2	building_id	smallint6	楼宇 ID
3	total_storey	smallint6	楼层数
4	total_area	double	总面积
5	height	double	楼高
6	type	smallint6	类型
7	status	varchar9	状态

(7) 物业消费信息表(community_use)用来保存物业消费信息，其结构如表 4-7 所示。

表 4-7　物业消费信息表结构

编 号	字段名称	数据结构	说 明
1	district_id	smallint6	小区 ID
2	date int11	int(11)	日期
3	tot_water_reading	double	用水量
4	tot_elec_reading	double	用电量
5	sec_supply_reading	double	用气量

注意：数据库逻辑结构设计的重要原则

在数据库设计过程中，数据库逻辑结构设计比较重要。概念结构是独立于任何一种数据模型的信息结构。逻辑结构设计的任务就是把概念结构设计阶段设计好的基本 E-R 图转换为与选用 DBMS 产品所支持的数据模型相符合的逻辑结构。在现实应用中，设计逻辑结构时一般分为以下三步。

(1) 将概念结构转换为一般的关系、网状和层次模型。
(2) 将转换后的关系、网状和层次模型向特定 DBMS 支持下的数据模型转换。
(3) 对数据模型进行优化。

数据库的概念结构和逻辑结构设计是数据库设计过程中最重要的两个环节。将概念模型转换为全局逻辑模型后，还应根据局部应用需求，结合具体 DBMS 的特点设计用户的外模式。当利用关系数据库管理系统的视图来完成外模式时，需要遵循以下设计原则。

(1) 使用符合用户习惯的别名。
(2) 针对不同级别的用户定义不同的外模式，以满足对安全性的要求。
(3) 简化用户对系统的使用，将经常使用的某些复杂查询定义为视图。

4.4 系统框架设计

系统框架设计步骤属于整个项目开发过程中的前期工作，项目中的具体功能将以此为基础进行扩展。经过细心的分析规划后，本项目需要通过以下 4 个阶段来实现。

扫码看视频

（1）搭建开发环境：操作系统为 Windows 7，数据库为 MySQL，开发工具为 Eclipse SDK。

（2）ADO 访问：建立 ADO 数据库访问连接类，是窗体能够访问数据库的前提。

（3）系统登录验证：确保合法用户才能登录系统。

（4）系统总体框架：此项目业务逻辑非常简单，所以各业务结点采用直接访问数据库的方式。

4.4.1 创建工程及设计主界面

1. 创建工程

（1）开发项目之前，要创建一个 Java 工程，用于管理以及调用整个工程中所用到的资源和代码。

（2）准备开发框架中所引用的第三方 jar 包和相应的技术文档。

（3）打开 IntelliJ IDEA，新建 Java 类型工程，弹出如图 4-4 所示的对话框。

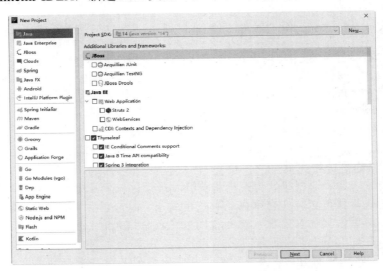

图 4-4　New Project 对话框

(4) 单击 Next 按钮,在弹出的对话框中设置 Project name(工程名称)为"JavaPrj_num4",然后自定义工程文件的保存路径,如图 4-5 所示。

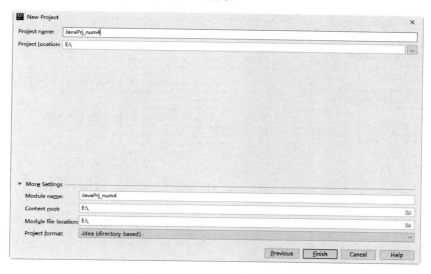

图 4-5 设置工程名和保存路径

(5) 如果在开发中需要引用第三方 jar 包,可以在 Project Structure 对话框中进行设置,如图 4-6 所示。

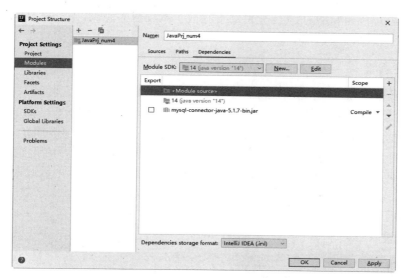

图 4-6 Project Structure 对话框

(6) 设置完成后，单击 Finish 按钮，即可创建 Java 工程，最终的工程目录结构如图 4-7 所示。

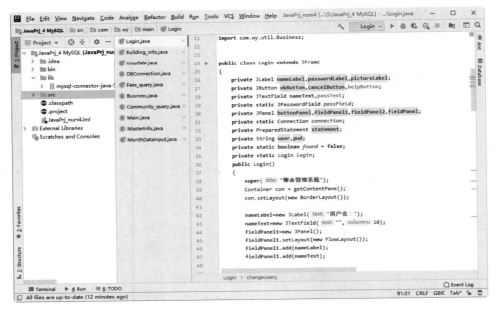

图 4-7 工程目录结构

(7) 在图 4-7 中，自动生成的 src 目录下存放了本项目的源码，在 lib 下显示了系统中引用的 jar 包。

2. 设计主界面

当用户登录成功后，会进入系统的主界面。主界面包括以下两大区域。

❑ 位于界面顶部的菜单栏，用于将系统所有功能进行归类展示。

❑ 位于菜单栏下面的工作区，是进行功能操作的区域，如图 4-8 所示。

(1) 主界面是整个系统通往各个功能模块的必经通道，因此要将各个功能模块的窗体加入主界面，同时要保证各窗体在主界面中布局合理，让用户方便操作。因此在主窗体中应加入整个系统的入口函数 main，通过执行该函数进而操作整个系统。主函数文件 Main.java 的实现代码如下：

```
public class Main extends JFrame
{       super("物业管理系统");
        Container con=getContentPane();
        //设置系统菜单以及其下的子菜单
        JMenuBar bar=new JMenuBar();
```

```
        setJMenuBar(bar);
        JMenu systemMenu=new JMenu("系统管理");
        JMenu infMenu=new JMenu("基本信息");
        JMenu enterMenu=new JMenu("消费指数");
        JMenu printMenu=new JMenu("各项费用报表");
}
```

在上述代码中,通过 super 函数设置了主窗体标题,然后为窗体增加菜单栏,并创建了多个 Jmenu 实例对象。

图 4-8　系统主界面

(2) 将为各一级菜单添加二级菜单,用于指向各功能模块的窗体。"退出系统"子菜单的定义代码如下。

```
JMenuItem exitItem=new JMenuItem("退出");
        exitItem.addActionListener(
            new ActionListener()
            {
                public void actionPerformed(ActionEvent event)
                {
                    System.exit(0);
                }
            }
        );
systemMenu.add(exitItem);//退出
bar.add(systemMenu);  //把 systemMenu 添加到菜单栏中
```

上述代码为子菜单 exitItem 添加了监听器 addActionListener,由 actionPerformed 函数的参数 event 识别该操作,最后执行 Java 系统函数 exit 退出系统。add 函数的功能是将创建的子菜单加入一级菜单中。

(3) 定义基本信息管理子菜单，定义代码如下。

```java
//设置物业信息维护菜单
JMenuItem placeItem=new JMenuItem("小区信息维护");
placeItem.addActionListener(new ActionListener()
{
    public void actionPerformed(ActionEvent event){
        Community_info xq = new Community_info();
    }
}
);
JMenuItem buildingItem=new JMenuItem("楼宇信息维护");
buildingItem.addActionListener(new ActionListener()
{
    public void actionPerformed(ActionEvent event){
        Building_info ly = new Building_info();
    }
}
);
JMenuItem hostItem=new JMenuItem("房屋信息维护");
hostItem.addActionListener(new ActionListener()
{
    public void actionPerformed(ActionEvent event){
        MasterInfo ui = new MasterInfo();
    }
}
);
infMenu.add(placeItem);
infMenu.add(buildingItem);
infMenu.add(hostItem);
JMenuItem enterPerPrice=new JMenuItem("修改收费单价");
enterPerPrice.addActionListener(new ActionListener()
{
    public void actionPerformed(ActionEvent event){
        PriceChange cp1 = new PriceChange();
    }
}
);
JMenuItem printPerPrice=new JMenuItem("查询收费单价");
printPerPrice.addActionListener(new ActionListener()
{
    public void actionPerformed(ActionEvent event){
        Payment sf1 = new Payment();
    }
}
);
```

```
infMenu.add(enterPerPrice);
infMenu.add(printPerPrice);
bar.add(infMenu);
```

上述代码为系统基本信息菜单增加了小区、楼宇、房屋、收费项目类别和各类别的单价功能菜单,并且分别为这些子菜单设置了窗体链接。这样当用户选择子菜单时会进入各个功能窗体,即可进行相应业务操作。

3. 设计菜单列表

因为本系统的具体功能都是通过操作菜单实现的,所以在本项目中必须实现菜单列表。

系统菜单需要根据系统的整体功能进行规划,然后针对各个需要的功能来增加对应的窗体类。在类 Main 中可进行菜单的管理,各个菜单的具体说明如表 4-8 所示。

表 4-8 菜单说明

菜单名称	ID 属性
小区信息维护	placeItem
楼宇信息维护	buildingItem
房屋信息维护	hostItem
修改收费单价	enterPerPrice
查询收费单价	printPerPrice
业主水/电/气指数录入	b1
公共水/电指数录入	b2
电费收费报表	a1
水费收费报表	a2
气费收费报表	a3
用户收费报表	a4
物业收费报表	a5

4.4.2 数据库 ADO 访问类

建立 com.wy.util 包,右击 new 命令后选择 class 选项,打开 New Java Class 窗口,设置 Name 为 Business,用于实现和数据库的连接。类 Business 的定义代码如下。

```java
public class Business {
    private static final String DBDRIVER = "com.mysql.jdbc.Driver" ;//驱动类类名
    private static final String DBURL = "jdbc:mysql://localhost:3306/wy";//连接URL
    private static final String DBUSER = "root" ;        //数据库用户名
```

```java
    private static final String DBPASSWORD = "root";        //数据库密码
    private static Connection conn = null;
//声明一个连接对象
    public static Connection getConnection(){

        try {
            Class.forName(DBDRIVER);                    //注册驱动
        conn = DriverManager.getConnection(DBURL,DBUSER,DBPASSWORD);//获得连接对象
        } catch (ClassNotFoundException e) {
            e.printStackTrace();
        } catch (SQLException e) {
            e.printStackTrace();
        }
        return conn;
    }
    public static ResultSet executeQuery(String sql) {    //执行查询方法
        try {
            if(conn==null)  new Business();    //如果连接对象为空,则重新调用构造方法
            return conn.createStatement(ResultSet.TYPE_SCROLL_SENSITIVE,
                    ResultSet.CONCUR_UPDATABLE).executeQuery(sql);//执行查询
        } catch (SQLException e) {
            e.printStackTrace();
            return null;        //返回null值
        } finally {
        }
    }
    public static int executeUpdate(String sql) {        //更新方法
        try {
            if(conn==null)  new Business();    //如果连接对象为空,则重新调用构造方法
            return conn.createStatement().executeUpdate(sql);//执行更新
        } catch (SQLException e) {
            e.printStackTrace();
            return -1;
        } finally {
        }
    }
    public static void close() {    //关闭方法
        try {
            conn.close();            //关闭连接对象
        } catch (SQLException e) {
            e.printStackTrace();
        }finally{
            conn = null;            //设置连接对象为null值
        }
    }
}
```

在上述代码中，Business 方法用于与数据库建立连接，在执行任何操作前都要调用此方法；executeQuery 方法用于执行查询并返回结果；executeUpdate 方法用于更新数据；close 方法用于断开与数据库的连接，在执行完任何访问数据库的相关操作后调用。

4.4.3 系统登录模块设计

为了增强系统的安全性，要求只有通过系统身份验证的用户才能使用本系统，为此必须增加一个系统登录功能模块。

(1) 添加 Login 类，定义成员变量，用来记录当前登录名和用户类型信息。

(2) Login 类继承于 JFrame 类，类 JFrame 是 Java 系统函数中窗体的基类。在登录窗体中添加 JLable 控件、JButton 控件和 JTextField 控件，代码如下。

```java
public class Login extends JFrame
{
    public Login()
    {
        super("物业管理系统");
        Container con = getContentPane();
        con.setLayout(new BorderLayout());
        nameLabel=new JLabel("用户名: ");
        nameText=new JTextField("",10);
        fieldPanel1=new JPanel();
        fieldPanel1.setLayout(new FlowLayout());
        fieldPanel1.add(nameLabel);
        passwordLabel = new JLabel(" 密 码: ");
        passField=new JPasswordField(10);
        fieldPanel2=new JPanel();
        fieldPanel2.setLayout(new FlowLayout());
        fieldPanel2.add(passwordLabel);
        fieldPanel2.add(passField);
        fieldPanel = new JPanel();
        fieldPanel.setLayout(new BorderLayout());
        fieldPanel.add(fieldPanel1,BorderLayout.NORTH);
        fieldPanel.add(fieldPanel2,BorderLayout.SOUTH);
        okButton=new JButton("确定");
        okButton.addActionListener(new LoginCheck());
        cancelButton = new JButton("取消");
        cancelButton.addActionListener(
            new ActionListener()
            {
                public void actionPerformed(ActionEvent event)
                {
                    System.exit(0);
                }
```

```
        }
    );
```

> **注意**：使用 MD5 加密技术加密登录信息

上述实现的是一个典型的登录验证模块，在软件项目中，这个模块具有代表性。对于本项目来说，因为使用的是 MySQL 数据库，所以系统一旦遭到黑客攻击，数据库信息被盗，登录信息就很容易被窃取。此时我们可以考虑采用 MD5 加密技术对系统进行升级。在用户设置登录密码时，在数据库中存储的便是 MD5 加密后的数据，即使黑客获取了数据库，打开后看到的也是加密信息，而不会得到正确的密码。

在上述代码中，函数 super 的功能是设置窗体标题；函数 getContentPane 的功能是刷新窗体，为添加组件做准备。同时，setLayout 对象用于进行窗口初始化布局，函数 add 用于将新建的组件加入 JPanel 组件中。窗体中的组件都是通过 Panel 组件进行统一管理的。

(3) 为窗体中的"确定"按钮增加监听函数 ActionListener，以实现用户的登录操作，其代码如下。

```java
private class LoginCheck implements ActionListener
{
    public void actionPerformed(ActionEvent event)
    {
        user = nameText.getText();        //获得用户名
        pwd = new String( passField.getPassword() );

        try{
            connection=Business.getConnection();      //建立连接
            String sql = "select * from user";
            statement = connection.prepareStatement(sql);
            ResultSet result = statement.executeQuery();
            while(result.next())
            {
if(user.compareTo(result.getString("uname"))==0&& pwd.compareTo
                (result.getString("paswrd")) == 0 ) {
                    found = true;
                    Main.mainFrame(user,result.getInt("purview"));
                    Login.this.setVisible(false);
                }
            }
            if( !found ){
                JOptionPane.showMessageDialog(null,"用户名或密码错误，请重输","错误",
                    JOptionPane.ERROR_MESSAGE);
                nameText.setText("");
                passField.setText("");
```

```
                }
                result.close();
                statement.close();
            }
            catch(Exception e) {
                e.printStackTrace();
            }
        }
    }
```

当输入"用户名"和"密码"后，系统会获取用户名和密码并提交到数据库进行查询。若系统中存在这个输入的用户名和密码，则启动主界面窗体 Main 的实例；否则提示输入错误，并弹出重新登录对话框。用户登录对话框如图 4-9 所示。

图 4-9　用户登录对话框

4.5　基本信息管理模块

在本物业管理系统中，基本信息管理是指对以下信息进行管理。
- 小区信息。
- 楼宇信息。
- 业主信息。
- 收费信息。
- 收费价格信息。

扫码看视频

4.5.1　小区信息管理

(1) 在项目中增加小区信息类 Community_info，此类继承于 Java 框架中的 JFrame 类，功能是新建一个窗体。因为在该窗体中不但可以增加小区信息，而且还要将结果显示在窗体中，所以要用表格布局展示结果。其定义代码如下：

```
public class Community_info extends JFrame{
    private JButton addButton = new JButton("添加");
    private JButton changeButton = new JButton("修改");
    private JButton deleteButton = new JButton("删除");
```

```
private JButton renewButton = new JButton("重置");
private JButton updateButton= new JButton("更新");
String title[]= {"小区编号","小区名称","小区地址","占地面积"};
Vector vector=new Vector();
Connection connection = null;
ResultSet rSet = null;
Statement statement = null;
AbstractTableModel tm;
public Community_info()
{
    enableEvents(AWTEvent.WINDOW_EVENT_MASK);
    try
    {
        jbInit();
    }
    catch(Exception e)
    {
        e.printStackTrace();
    }
}
```

在上述代码中，集中实现了"添加""修改""删除""重置"和"更新"按钮，并且增加了相应的监听器，以便数据库对各种操作做出响应。其中，enableEvents 函数的功能是触发窗体表格屏蔽事件。

(2) 表格布局实现过程的定义代码如下。

```
private void createtable()       //创建小区信息表格
{
    JTable table;                //表格
    JScrollPane scroll;          //滚动条
    tm = new AbstractTableModel() {
        public int getColumnCount() {
            return title.length;
        }
        public int getRowCount() {
            return vector.size();
        }
        public String getColumnName(int col) {
            return title[col];
        }
        public Object getValueAt(int row, int column) {
            if (!vector.isEmpty()) {
                return ((Vector) vector.elementAt(row)).elementAt(column);
            } else {
                return null;
            }
```

```
            }
            public void setValueAt(Object value, int row, int column) {
            }
            public Class getColumnClass(int c) {
                return getValueAt(0, c).getClass();
            }
            public boolean isCellEditable(int row, int column) {
                return false;
            }
        };
        table = new JTable(tm);
        table.setToolTipText("Display Query Result");
        table.setAutoResizeMode(table.AUTO_RESIZE_OFF);
        table.setCellSelectionEnabled(false);
        table.setShowHorizontalLines(true);
        table.setShowVerticalLines(true);
        scroll = new JScrollPane(table);
        scroll.setBounds(6,20,540,250) ;
        tablePanel.add(scroll);
    }
```

在上述代码中，用到了类 AbstractTableModel。Java 提供的 AbstractTableModel 是一个抽象类，这个类能帮助我们实现大部分的 TableModel 方法，除了 getRowCount()、getColumnCount()、getValueAt()三个方法。因此，我们的主要任务是实现这三个方法，然后就可以利用这个抽象类设计出不同格式的表格。

> **注意：抽象类 AbstractTableModel 中的方法**
>
> AbstractTableModel 类提供了以下方法。
> - void addTableModelListener(TableModelListener l): 使表格具有处理 TableModelEvent 的能力。当表格的 Table Model 发生变化时，会发送 TableModelEvent 信息。
> - int findColumn(String columnName): 在列名称中寻找 columnName 这个项目。若有，则返回其所在列的位置；反之则返回-1，表示未找到。
> - void fireTableCellUpdated(int row, int column): 通知所有的 Listener(监听器)，这个表格中(row,column)字段的内容已经改变。
> - void fireTableChanged(TableModelEvent e): 将所接收的事件通知传送给所有在这个 Table Model 中注册过的 TableModelListeners。
> - void fireTableDataChanged(): 通知所有的监听器，在这个表格中列的内容已经改变了，列的数目也可能已经改变了。因此 JTable 可能需要重新显示此表格的结构。
> - void fireTableRowsDeleted(int firstRow, int lastRow): 通知所有的监听器，在这个表格中，第 firstrow 行至 lastrow 行数据已经被删除了。

- void fireTableRowsUpdated(int firstRow, int lastRow)：通知所有的监听器，在这个表格中，第 firstrow 行至 lastrow 行数据已经被修改了。
- void fireTableRowsInserted(int firstRow, int lastRow)：通知所有的监听器，在这个表格中，第 firstrow 行至 lastrow 行已经被插入数据。
- void fireTableStructureChanged()：通知所有的监听器，在这个表格的结构已经改变了，包括行的数目、名称以及数据类型都可能已经改变了。
- Class getColumnClass(int columnIndex)：返回字段数据类型的类名称。
- String getColumnName(int column)：若没有设置列标题，则返回默认值，依次为 A,B,C,…,Z,AA,AB,…。若无此 column，则返回一个空的字符串。
- Public EventListener[] getListeners(Class listenerType)：返回所有在这个 Table Model 中建立的符合 listenerType 的监听器，并以数组形式返回。
- boolean isCellEditable(int rowIndex, int columnIndex)：返回所有在这个 Table Model 中建立的符合 listenerType 形式的监听器，并以数组形式返回。
- void removeTableModelListener(TableModelListener l)：从 TableModelListener 中移除一个监听器。
- void setValueAt(Object aValue, int rowIndex, int columnIndex)：设置 cell(rowIndex, columnIndex)单元格的值。

(3) 创建一个名为 addButton_actionPerformed 的方法，这是一个用于"添加"按钮的监听器实现方法。在该方法中与数据库建立连接后，将窗体中各输入组件的值赋值给变量，并通过执行 SQL 语句将信息插入小区信息表中。其代码如下：

```
void addButton_actionPerformed(ActionEvent e)          //添加按钮事件方法
    {
        try{
            connection=Business.getConnection();    //建立连接
            statement = connection.createStatement();
            String sql1=
"insert into community_info(district_id,district_name,address,floor_space)
values("+Integer.parseInt(districtid.getText())+",'"+buildingid.getText()+"',
'"+stac.getText()+"',"+Integer.parseInt(func.getText())+")";
            statement.executeUpdate(sql1);          //执行插入操作
        }
        catch(Exception ex){
            JOptionPane.showMessageDialog(Community_info.this,"添加数据出错",
                                    "错误",JOptionPane.ERROR_MESSAGE);
        }
        finally{
            try{
                if(statement != null){
```

```
                    statement.close();
                }
                if(connection != null){
                    connection.close();
                }
            }
            catch(SQLException ex){
            }
        }
        selectDistrict();
    }
    void changeButton_actionPerformed(ActionEvent e);    //修改监听器
    void deleteButton_actionPerformed(ActionEvent e);    //删除监听器
    void updateButton_actionPerformed(ActionEvent e);    //更新监听器
```

在上述代码中，为窗体中的其他按钮增加了监听器操作，小区信息维护界面的运行效果如图 4-10 所示。

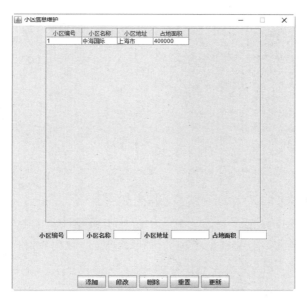

图 4-10 小区信息维护界面

4.5.2 楼宇信息管理

(1) 在工程中创建新增楼宇信息类 Building_info，该类继承于 Java 框架中的 JFrame 类。因为在该窗体中不但可以增加楼宇信息，而且增加后要将结果显示在窗体上，所以需要在窗体上进行表格布局以展示处理结果。其代码如下。

```java
public class Building_info extends JFrame
{
private JButton searchButton = new JButton("按小区查询");
    private JButton addButton = new JButton("添加");
    private JButton changeButton = new JButton("修改");
    private JButton deleteButton = new JButton("删除");
    private JButton renewButton = new JButton("重置");
    private JButton updateButton= new JButton("更新");
     public Building_info(){
        enableEvents(AWTEvent.WINDOW_EVENT_MASK);
        try {
            jbInit();
        }
        catch(Exception e)
        {
            e.printStackTrace();
        }
    }
    private void jbInit() throws Exception
    {
        Container con = getContentPane();
        con.setLayout(new BorderLayout());
        label1.setText("小区编号");
        label2.setText("楼宇编号");
        label3.setText("楼宇层数");
        label4.setText("产权面积");
        label5.setText("楼宇高度");
        label6.setText("类型");
        label7.setText("楼宇状态");
    }
}
```

在上述代码中，集中实现了"添加""修改""删除""重置"和"更新"按钮，并增加了相应的监听器操作，能够根据各种操作执行相应的数据库操作。函数 getContentPane 能够刷新窗体，初始化窗体组件，为新加入的组件做准备。

（2）添加监听器的实现方法，其代码如下。

```java
void addButton_actionPerformed(ActionEvent e)
    {
        try{
            connection=Business.getConnection();            //建立连接
            statement = connection.createStatement();       //创建SQL会话
            String sql1= "insert into building_info(district_id,building_id,total_storey,total_area,height,type,status)values("+Integer.parseInt(districtid.getText())+","+Integer.parseInt(buildingid.getText())+","+Integer.parseInt(storey.getText())+",
```

```
"+Integer.parseInt(area.getText())+","+Integer.parseInt(height.getText())+","+
Integer.parseInt(type.getText())+",'"+state.getText()+"')";
            statement.executeUpdate(sql1);
        }
        catch(Exception ex){
            JOptionPane.showMessageDialog(Building_info.this,"添加数据出错",
                                "错误",JOptionPane.ERROR_MESSAGE);
        }
        finally{
            try{
                if(statement != null){
                    statement.close();
                }
                if(connection != null){
                    connection.close();
                }
            }
            catch(SQLException ex){
            }
        }
        selectBuilding();
    }
```

在上述代码中,为窗体中的其他按钮实现监听器操作,主要包括 searchButton_actionPerformed(查询)、updateButton_actionPerformed(更新)、deleteButton_actionPerformed(删除)和 changeButton_actionPerformed(修改)监听器。最终楼宇信息管理增加的运行效果如图 4-11 所示。

注意:insert into 和 replace into 的区别

在上述代码中使用了 insert into 语句,这个语句的功能与 replace into 类似,但有一个例外:如果表中的一个旧记录与新记录具有相同的值,并且这个字段是用于 PRIMARY KEY 或 UNIQUE 索引的,那么在插入新记录之前,旧记录将被删除。如果数据表中没有 PRIMARY KEY 或 UNIQUE 索引,那么使用 replace 语句将没有意义,因为它与 insert 的行为相同,没有索引用于确定新行是否复制了其他行。

所有列的值均取自 replace 语句中被指定的值,所有缺失的列被设置为各自的默认值,这和 insert 一样。不能从当前行中引用值,也不能在新行中使用值。如果使用一个例如 "SET col_name = col_name + 1" 的赋值,则位于右侧的列名称的引用会被作为 DEFAULT(col_name)处理。因此,该赋值相当于 "SET col_name = DEFAULT(col_name) + 1"。

为了能够使用 replace,必须同时拥有表的 insert 和 delete 权限。replace 语句会返回一个数,表示受影响的行的数目,该数是被删除和被插入的行数的和。如果该数为 1,则一行被插入，同时没有行被删除；如果该数大于1，则在新行被插入前，有一个或多个旧行被删除。如果表包含多个唯一索引，并且新行复制了不同唯一索引中的不同旧行的值，则有可能是一行替换了多个旧行。

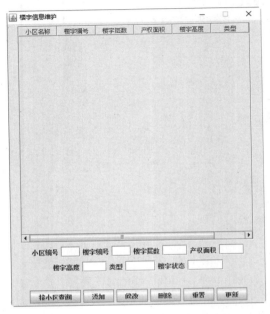

图 4-11　楼宇信息维护界面

4.5.3　业主信息管理

（1）在工程中创建添加业主信息类 MasterInfo，该类继承于 Java 框架中的 JFrame 类。在该窗体中不但可以增加业主信息，而且窗体需要采用表格布局的样式展示结果。在类 MasterInfo 的构造函数中实现窗体初始化操作，并利用 jbInit 函数进行组件布局管理。其代码如下。

```java
public MasterInfo() {
    enableEvents(AWTEvent.WINDOW_EVENT_MASK);
    try {
        jbInit();              //调用初始化方法
    } catch (Exception e) {
        e.printStackTrace();
    }
    setSize(480, 600);
    setResizable(false);
    setVisible(true);
```

```java
    }
    private void jbInit() throws Exception {
        Container con = getContentPane();
        con.setLayout(new BorderLayout());
        label1.setText("小区");
        label2.setText("楼号");
        label3.setText("房号");
        label4.setText("产权面积");
        label5.setText("房屋状态");
        label6.setText("用途");
        label7.setText("姓名");
        label8.setText("性别");
        label9.setText("身份证");
        label10.setText("联系地址");
        label11.setText("联系电话");
    setTitle("业主信息维护");
```

在上述代码中，通过jbInit函数集中实现了"添加""删除""修改""重置"和"更新"按钮，增加了相应的监听器操作，并初始化窗体组件，为新加入组件做准备。

(2) 添加监听器的实现方法，其定义代码如下。

```java
void deleteButton_actionPerformed(ActionEvent e) {           //删除按钮事件方法
        try {
            connection = Business.getConnection();
            statement = connection.createStatement();
            String sql2 = "delete from room_info where district_id="
                + Integer.parseInt(districtid.getText())
                + "AND building_id="
                + Integer.parseInt(buildingid.getText()) + "AND room_id="
                + Integer.parseInt(roomid.getText());// +"area="+Integer.parseInt(area.getText())+"||oname="+ownername.getText()+"||sex="+sex.getText()+"||id_num="+idnum.getText()+"||address="+addr.getText()+"||phone="+Integer.parseInt(tetel.getText());
            statement.executeUpdate(sql2);

        } catch (Exception ex) {
            JOptionPane.showMessageDialog(MasterInfo.this, "删除数据出错", "错误",
                JOptionPane.ERROR_MESSAGE);
        } finally {
            try {
                if (statement != null) {
                    statement.close();
                }
                if (connection != null) {
                    connection.close();
                }
            } catch (SQLException ex) {
```

```
            }
        }
        selectUser();
}
```

在上述代码中，为窗体中的其他按钮增加了监听器操作，主要包括 searchButton_actionPerformed(查询)、updateButton_actionPerformed(更新)、addButton_actionPerformed(添加)和 changeButton_actionPerformed(修改)监听器。业主信息维护界面的运行效果如图 4-12 所示。

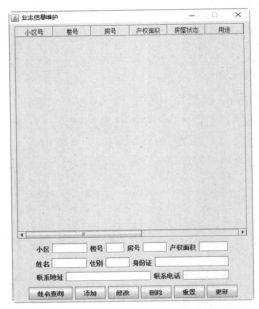

图 4-12　业主信息维护界面

4.5.4　收费信息管理

(1) 在工程中创建收费信息维护类 PriceChange，该类继承于 Java 框架的 JFrame 类。在类 PriceChange 中定义了该窗体需要的各种组件，包括文本输入框、JLable 标签、JButton 按钮、Panel 容器等组件。类 PriceChange 采用 GridLayout 进行布局管理，其代码如下。

```
public PriceChange() {
    try {
        c1 = Business.getConnection();              //建立连接
        String sql = "SELECT * FROM price_type";    //查询收费信息
        stmt1 = c1.createStatement();
```

```
            rsl1 = stmt1.executeQuery(sql);          //执行 SQL 查询
            rsmd1 = rsl1.getMetaData();              //获取表的字段名称
        } catch (Exception err) {
            err.printStackTrace();
        }
        JPanel button_con = new JPanel();
        button_con.setLayout(new FlowLayout());
        comm_1 = new JButton("增加");
        comm_1.addActionListener(new ActionListener() {
            public void actionPerformed(ActionEvent e) {
                addPrice();
            }
        });
        comm_2 = new JButton("修改");
        comm_2.addActionListener(new ActionListener() {
            public void actionPerformed(ActionEvent e) {
                changePrice();
            }
        });
        comm_3 = new JButton("删除");
        comm_3.addActionListener(new ActionListener() {
            public void actionPerformed(ActionEvent e) {
                deletePrice();
            }
        });
    }
```

在上述代码中分别创建了"增加""修改"和"删除"按钮,并增加了相应的监听器。

(2) 类 GridLayout 是一个布局处理器,它是以矩形网格形式对容器的组件进行布置。GridLayout 容器被分成大小相等的矩形,在每个矩形中放置一个组件,利用表格的形式展示信息。其代码如下。

```
public void getpriceTable() {
    try {
        Vector columnName = new Vector();                //创建表头
        Vector rows = new Vector();                      //表格信息集合
        columnName.addElement("收费项目编号");              //添加信息
        columnName.addElement("收费项目名称");
        columnName.addElement("单价(单位:元)");
        while (rsl1.next()) {                            //遍历结果集
            Vector currentRow = new Vector();            //创建收费信息集合
            for (int i = 1; i <= rsmd1.getColumnCount(); ++i) {
                currentRow.addElement(rsl1.getString(i));    //添加信息
            }
            rows.addElement(currentRow);                 //保存到集合中
        }
```

```
            display = new JTable(rows, columnName);          //创建表格
            JScrollPane scroller = new JScrollPane(display);
            Container temp = getContentPane();
            temp.add(scroller, BorderLayout.CENTER);
            temp.validate();
        } catch (SQLException sqlex) {
            sqlex.printStackTrace();
        }
    }
```

在上述代码中，为窗体中的其他按钮增加了监听器操作，主要包括 addPrice(增加)、changePrice(修改)、deletePrice(删除)等监听器。收费信息维护界面的运行效果如图 4-13 所示。

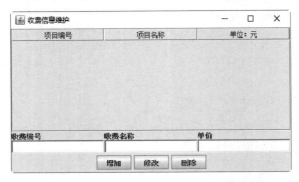

图 4-13　收费信息维护界面

注意：充分认识 JTable 组件的不足

在上述代码中用到了 Swing 中的 JTable 组件。在 JTable 中有以下两个接收数据的 JTable 构造器。

- JTable(Object[][] rowData, Object[] columnNames)
- JTable(Vector rowData, Vector columnNames)

上述构造函数的好处是容易实现，但缺点也十分明显，具体说明如下。

- 会自动设置每个单元格为可编辑。
- 将数据类型都视为一样的(字符串类型)。例如，如果表格的一列有 Boolean 数据，则表格用单选框来展示这个数据。如果用上面两种构造方法，则 Boolean 数据将显示为字符串。
- 要求把所有表格数据放入一个数组或者 Vector 中，这些数据结构可能不适合某些数据类型。

4.5.5 查询单价清单

在工程中创建收费单价清单类 Payment，该类继承于 Java 框架中的 JFrame 类，其中定义了该窗体所需要的各种组件。收费单价清单窗体采用表格布局的方式展示了清单信息，在实现时首先创建了表格，为表格增加表头；然后循环取出数据源中的数据，并写入表格的每一行中。其代码如下。

```java
public class Payment extends JFrame {
public void getpriceTable(){
    try {
            Vector columnName = new Vector();   //显示列名
            Vector rows = new Vector();//定义向量，进行数据库数据操作
            String sql="SELECT * FROM price_type";
            Statement stmt1 = c1.createStatement();
            ResultSet rsl1 = stmt1.executeQuery(sql);
            ResultSetMetaData rsmd1 = rsl1.getMetaData();//获取表的字段名称
            columnName.addElement("收费项目");
            columnName.addElement("单位:元");
            while(rsl1.next())
            {                                           // 获取记录集
                Vector currentRow = new Vector();
                for ( int i = 2; i <= rsmd1.getColumnCount(); ++i )
                {
                    currentRow.addElement( rsl1.getString( i ) );
                }
                rows.addElement(currentRow);
            }
            display = new JTable(rows,columnName);
            JScrollPane scroller = new JScrollPane(display);
            Container temp = getContentPane();
            temp.add(scroller,BorderLayout.CENTER);
            temp.validate();
            }
            catch(SQLException sqlex)
            {
                sqlex.printStackTrace();
            }
        }
    }
```

在上述代码中，实现了 JTable 函数中的两个参数，建立了一个以 Vector 为输入来源的数据表格，可以显示列的名称。函数 addElement 用于将每一列加入表格中。

收费单价清单界面的运行效果如图 4-14 所示。

图 4-14 收费单价清单界面

在上述代码中，Vector 中的 add 方法与 addElement 方法是有区别的。Add 方法能够将指定元素添加到此向量的末尾，主要针对元素进行操作。addElement 方法能够将指定的组件添加到此向量的末尾，将其大小增加 1，主要针对组件进行操作。

4.6 消费指数管理模块

根据系统分析规划书，本项目的系统消费指数分为以下两类。
- 业主：水、电、气录入。
- 物业：水、电、气录入。

扫码看视频

4.6.1 业主消费录入

（1）在工程中创建添加业主录入类 MonthDataInput，并且定义该窗体需要的各种组件，如 Panel 容器和下拉列表框等组件。在录入业主消费指数操作中，先选择对应的楼宇和小区信息，然后进入另外一个界面中进行录入。其代码如下：

```java
public MonthDataInput(int tt) {   //不带参数的界面
        super("选择小区和楼宇");
        type = tt;
        mainperform(type);
        time = 0;
        t = 0;
}
    public MonthDataInput(String uptown, String uptown_id, int tt) {
        super("选择小区和楼宇");
        type = tt;
        mainperform(uptown, uptown_id, type);
```

```
        time = 0;
    }
    public MonthDataInput(String uptown, String building, String uptown_id) {
        super("输入数据");
        mainperform(uptown, building, uptown_id);
        time = 0;
    }
```

在上述代码中定义了三个构造函数,其中,第一个为窗体默认构造函数,主要用于初始化;第二个为物业消费指数添加界面,只针对物业;第三个为业主消费指数添加界面,主要是针对业主,这种分类标识的模式更易管理。业主消费录入模块的界面效果如图 4-15 所示。

图 4-15　业主消费录入模块界面

(2) 以下是 mainperform 构造函数的具体实现,仍然用 GridBagLayout 进行布局管理。用户单击下拉表框后,会根据所选参数进入不同的界面。其代码如下。

```
public void mainperform(String inuptown, String inuptownid, int tt)
    {
        time++;
        Container panelin = getContentPane();
        gridbag = new GridBagLayout();
        panelin.setLayout(gridbag);
        uptownid = new String[100];
        uptownname = new String[100];
        uptownname[0] = new String(inuptown);
        getuptown();
        type = tt;
        int i = 0;
        do {
            i++;
        } while (!uptownname[i].equals(inuptown));
        buildingid = new String[150];
```

```
        getbuilding(inuptownid);
        buildingid[0] = new String("选择楼宇");
        roomid = new String[150];
        uptown = new String(inuptown);
        uptownid_select = new String(inuptownid);
        uptownlabel = new JLabel("选择小区");
        uptownlabel.setToolTipText("点击选择小区");
        inset = new Insets(5, 5, 5, 5);
        c = new GridBagConstraints(2, 1, 1, 1, 0, 0, 10, 0, inset, 0, 0);
        gridbag.setConstraints(uptownlabel, c);
        panelin.add(uptownlabel);
        uptownselect = new JComboBox(uptownname);
        uptownselect.setSelectedIndex(i);
        uptownselect.setMaximumRowCount(5);
        uptownselect.addItemListener(new ItemListener() {
            public void itemStateChanged(ItemEvent event) {
                int i = 0;
                i = uptownselect.getSelectedIndex();
                t++;
                getbuilding(uptownid[i]);
                if (t < 2) {
                    MonthDataInput mdif2 = new MonthDataInput(
                            uptownname[i], uptownid[i], type);
                }
                MonthDataInput.this.setVisible(false);
            }
        });
        c = new GridBagConstraints(5, 4, 1, 2, 0, 0, 10, 0, inset, 0, 0);
        gridbag.setConstraints(button2, c);
```

> **注意：GridBagLayout 窗体布局方式**

在上述代码中，使用了 GridBagLayout 窗体布局的知识。java.awt.GridBagLayout 是一个灵活的布局管理器，它不要求组件的大小相同便可以将组件垂直、水平或沿它们的基线对齐。每个 GridBagLayout 对象维持一个动态的矩形单元网格，每个组件占用一个或多个这样的单元，该单元被称为显示区域。每个由 GridBagLayout 管理的组件都与 GridBagConstraints 的实例相关联。Constraints 对象指定组件的显示区域在网格中的具体位置，以及组件在其显示区域中的放置方式。除了 Constraints 对象之外，GridBagLayout 还考虑每个组件的最小大小和首选大小，以确定组件的大小。网格的总体方向取决于容器的 ComponentOrientation 属性。对于水平的从左到右的方向，网格坐标(0,0)位于容器的左上角，其中，X 向右递增，Y 向下递增。对于水平的从右到左的方向，网格坐标(0,0)位于容器的右上角，其中，X 向左递增，Y 向下递增。

要想使用 GridBagLayout，必须使用 GridBagConstraints 对象来指定 GridBagLayout 中组件的位置。类 GridBagLayout 的 setConstraints 方法用 Component 和 GridBagConstraints 作为参数来设置组件的约束。

(3) 在本项目中，业主消费量录入窗体主要包括用水量、用电量和用气量的录入，同时必须输入日期。按照上述流程完成每月业主消费指数的录入后，数据供费用报表引用。业主消费功能由类 ChargeReport 实现，主要实现代码如下。

```java
public class ChargeReport extends JFrame {

    private JPanel buttonPanel, tablePanel, fieldPanel, fieldPanel1,
            fieldPanel2, fieldPanel3;

    String title[] = { "小区号", "楼号", "房号", "业主姓名", "用途", "上月读数",
            "本月读数", "单价", "应缴总额" };              //表头信息

    Vector vector = new Vector();

    Connection connection = null;
    ResultSet rSet = null;
    Statement statement = null;
    AbstractTableModel tm;

    private int tt, did, bid, time;
    private double uPrice;

    public ChargeReport(String type, int district_id, int building_id) {
        super(type);
        if (type.compareTo("电费收费报表") == 0)         //判断是查询什么报表
            tt = 1;
        else if (type.compareTo("天燃气费收费报表") == 0)
            tt = 2;
        else if (type.compareTo("水费收费报表") == 0)
            tt = 3;

        did = district_id;
        bid = building_id;
        time = nowdate.getSystime();

        enableEvents(AWTEvent.WINDOW_EVENT_MASK);
        try {
            jbInit();
        } catch (Exception e) {
            e.printStackTrace();
        }
```

```java
        setSize(480, 480);
        setResizable(false);
        setVisible(true);
    }

    private void jbInit() throws Exception {
        Container con = getContentPane();
        tablePanel = new JPanel();
        createtable();
        con.add(tablePanel);
        updated();
    }

    private void createtable() {
        JTable table;
        JScrollPane scroll;
        tm = new AbstractTableModel() {
            public int getColumnCount() {
                return title.length;
            }

            public int getRowCount() {
                return vector.size();
            }

            public String getColumnName(int col) {
                return title[col];
            }

            public Object getValueAt(int row, int column) {
                if (!vector.isEmpty()) {
                    return ((Vector) vector.elementAt(row)).elementAt(column);
                } else {
                    return null;
                }

            }

            public void setValueAt(Object value, int row, int column) {

            }

            public Class getColumnClass(int c) {
                return getValueAt(0, c).getClass();
            }

            public boolean isCellEditable(int row, int column) {
```

```java
                return false;
            }
        };

        table = new JTable(tm);
        table.setToolTipText("Display Query Result");
        table.setAutoResizeMode(table.AUTO_RESIZE_OFF);
        table.setCellSelectionEnabled(false);
        table.setShowHorizontalLines(true);
        table.setShowVerticalLines(true);
        scroll = new JScrollPane(table);
        scroll.setBounds(6, 20, 540, 250);
        tablePanel.add(scroll);
    }
    void updated() {                    //执行查询
        getUnitPrice();
        try {
            connection = Business.getConnection();
            statement = connection.createStatement();
            String sql3 = null;
            sql3 = "select r.district_id,r.building_id,r.room_id,r.oname,
r.purpose,ur1.water_reading,ur2.water_reading from room_info r,user_reading ur1,
user_reading ur2 where r.district_id=" + did + " and r.district_id=ur1.district_id
and ur1.district_id=ur2.district_id and r.building_id=" + bid + " and r.building_id=
ur1.building_id and ur1.building_id= ur2.building_id and r.room_id=ur1.room_id and
ur1.room_id=ur2.room_id and ur1.date="
                    + (time - 1) + " and ur2.date=" + time;     //查询SQL语句
            rSet = statement.executeQuery(sql3);
            if (rSet.next() == true) {
                String sql = sql3;
                ResultSet rs = statement.executeQuery(sql);
                vector.removeAllElements();
                tm.fireTableStructureChanged();
                while (rs.next()) {              //对结果集进行遍历
                    Vector rec_vector = new Vector();  //信息集合
                    //小区号
                    rec_vector.addElement(String.valueOf(rs.getInt(1)));
                    //楼号
                    rec_vector.addElement(String.valueOf(rs.getInt(2)));
                    //房号
                    rec_vector.addElement(String.valueOf(rs.getInt(3)));
                    rec_vector.addElement(rs.getString(4));      //业主姓名
                    rec_vector.addElement(rs.getString(5));      //用途
                    double last = rs.getDouble(6);               //上月读数
                    double current = rs.getDouble(7);            //本月读数
                    rec_vector.addElement(String.valueOf(last));
                    rec_vector.addElement(String.valueOf(current));
```

```java
                    rec_vector.addElement(String.valueOf(uPrice));
                    rec_vector.addElement(String.valueOf((current - last)
                            * uPrice));          //计算金额
                    vector.addElement(rec_vector);
                }
                tm.fireTableStructureChanged();
            }
        } catch (Exception ex) {
            ex.printStackTrace();
        } finally {
            try {
                if (statement != null) {
                    statement.close();
                }
                if (connection != null) {
                    connection.close();
                }
            } catch (SQLException ex) {
                System.out.println("\nERROR:---------SQLException--------\n");
                System.out.println("Message: " + ex.getMessage());
                System.out.println("SQLState: " + ex.getSQLState());
                System.out.println("ErrorCode: " + ex.getErrorCode());
            }
        }
    }
    void getUnitPrice() {          //获取收费项目单价
        String sql = null;
        try {
            connection = Business.getConnection();
            statement = connection.createStatement();
            //查询指定项目单价
            sql = "select * from price_type where charge_id=" + tt;
            rSet = statement.executeQuery(sql);
            if (rSet.next() == true)
                uPrice = rSet.getDouble("unit_price");
        } catch (Exception ex) {
            ex.printStackTrace();
        } finally {
            try {
                if (statement != null) {
                    statement.close();
                }
                if (connection != null) {
                    connection.close();
                }
            } catch (SQLException ex) {
                System.out.println("\nERROR:---------SQLException--------\n");
```

```
                System.out.println("Message: " + ex.getMessage());
                System.out.println("SQLState: " + ex.getSQLState());
                System.out.println("ErrorCode: " + ex.getErrorCode());
            }
        }
    }
}
```

(4) 在业主消费录入窗体中用到了日期相关的函数,利用 nowdate 类可简化日期操作。因为 Java 提供的 Calendar 对象不是本系统所要求的格式,所以需要用 getInstance 函数进行转化,然后返回转化结果。其代码如下。

```
public class nowdate {
    public static int getSystime() {
        Calendar tt = Calendar.getInstance();          //获取 Calendar 对象
        Date time = tt.getTime();                      //获取日期
        int year = time.getYear() + 1900;              //获取年份
        int month = time.getMonth() + 1;               //获取月份
        return year * 100 + month;                     //组成数字
    }
    public static void main(String args[]) {
        Calendar tt = Calendar.getInstance();
        Date time = tt.getTime();
        int year = time.getYear() + 1900;
        int month = time.getMonth() + 1;
    }
}
```

4.6.2 物业消费录入

物业消费录入是物业公共区域消费指数的信息,只包含电量数据,无水和气之类的消费。物业消费录入同样由录入员每月进行录入,集中管理。物业消费录入功能的实现过程和业主消费录入功能类似,为节省篇幅,不再列出。

4.7 各项费用管理模块

根据系统需求分析可知,在本系统中包含以下两大类费用。
- 业主:包括水、电、气单项费用查询和总费用查询。
- 物业:物业费用查询。

扫码看视频

4.7.1 业主费用查询

(1) 在工程中创建添加费用查询类 Fees_query。考虑到功能类似,因此将这几种查询集中存于此类中,以不同的参数加以区别。其代码如下。

```java
public Fees_query(String uptownid, String uptownname, String buildingid,
        String roomid, int date) {
    JLabel titleLabel, waterLabel, eleLabel, gasLabel;
    JTextField waterField, eleField, gasField;
    JButton button1, button2;
    Container panelin = getContentPane();
    gridbag = new GridBagLayout();
    panelin.setLayout(gridbag);
    gridbag = new GridBagLayout();
    panelin.setLayout(gridbag);
    try {
        Class.forName("sun.jdbc.odbc.JdbcOdbcDriver");   //建立连接
        String url4 = "jdbc:odbc:estate";
        Connection connection4 = DriverManager.getConnection(url4);
        Statement stmt4 = connection4.createStatement();
        String sqlLastData = "SELECT water_reading, elec_reading,
                    gas_reading FROM master_use WHERE district_id="
            + uptownid
            + " AND building_id="
            + buildingid
            + " AND room_id=" + roomid + " AND date=" + date;
        ResultSet rsLastData = stmt4.executeQuery(sqlLastData);
        while (rsLastData.next()) {
            water = rsLastData.getString("water_reading");
            ele = rsLastData.getString("elec_reading");
            gas = rsLastData.getString("gas_reading");
        }
        rsLastData.close();
        connection4.close();
    }
    catch (Exception ex) {
        System.out.println(ex);
    }
```

在以上代码中,根据传递的参数从数据库中获得业主使用各种项目的指数和费用,如业主 ID 号、楼宇号、小区号和日期。这样可以方便物业管理员实时查看,避免了繁杂的纸质记录的麻烦。

(2) 获得数据后,通过表格布局展示结果,其代码如下。

```
titleLabel = new JLabel(uptownname + "中" + buildingid + "号楼 " + roomid
            + "在 " + date + "水表、电表和天燃气读数");
inset = new Insets(5, 5, 5, 5);
c = new GridBagConstraints(2, 1, 5, 1, 0, 0, 10, 0, inset, 0, 0);
gridbag.setConstraints(titleLabel, c);
panelin.add(titleLabel);
waterLabel = new JLabel("用水");
c = new GridBagConstraints(2, 5, 1, 1, 0, 0, 10, 0, inset, 0, 0);
gridbag.setConstraints(waterLabel, c);
panelin.add(waterLabel);
waterField = new JTextField(water, 7);
waterField.setEditable(false);
c = new GridBagConstraints(4, 5, 1, 1, 0, 0, 10, 0, inset, 0, 0);
gridbag.setConstraints(waterField, c);
panelin.add(waterField);
eleLabel = new JLabel("用电");
c = new GridBagConstraints(6, 5, 1, 1, 0, 0, 10, 0, inset, 0, 0);
gridbag.setConstraints(eleLabel, c);
panelin.add(eleLabel);
eleField = new JTextField(ele, 7);
eleField.setEditable(false);
c = new GridBagConstraints(8, 5, 1, 1, 0, 0, 10, 0, inset, 0, 0);
gridbag.setConstraints(eleField, c);
panelin.add(eleField)
gasLabel = new JLabel("气");
c = new GridBagConstraints(10, 5, 1, 1, 0, 0, 10, 0, inset, 0, 0);
gridbag.setConstraints(gasLabel, c);
panelin.add(gasLabel);

gasField = new JTextField(gas, 7);
gasField.setEditable(false);
c = new GridBagConstraints(12, 5, 1, 1, 0, 0, 10, 0, inset, 0, 0);
gridbag.setConstraints(gasField, c);
panelin.add(gasField);
```

在以上的代码中，运用 GridBagConstraints 实现了表格布局管理。当用户触发查询操作后，从数据库中取出相应的数据。因为水、电、气是业主每月都应交的费用，所以在同一个界面中显示，这样给使用者带来了方便。

（3）下面是查询操作的监听器定义，其代码如下。

```
button1 = new JButton("查询");
button1.addActionListener(new ActionListener() {
    public void actionPerformed(ActionEvent event) {
        Fees_query rdif = new Fees_query(3);
        Fees_query.this.setVisible(false);
    }
```

```
        });
        c = new GridBagConstraints(5, 7, 2, 2, 0, 0, 10, 0, inset, 0, 0);
        gridbag.setConstraints(button1, c);
        panelin.add(button1);
        button2 = new JButton("返回");
        button2.addActionListener(new ActionListener() {
            public void actionPerformed(ActionEvent event) {
                Fees_query.this.setVisible(false);
            }
        });
        c = new GridBagConstraints(8, 7, 2, 2, 0, 0, 10, 0, inset, 0, 0);
        gridbag.setConstraints(button2, c);
        panelin.add(button2);
        setSize(800, 300);
        setVisible(true);
    }
```

4.7.2 物业费用查询

(1) 在类 Fees_query 中增加了与业主查询相类似的功能。物业查询包括按小区查询、全部记录查询和楼宇查询功能,将这几种查询集中在类 Fees_query 中实现,以不同的参数加以区别。其代码如下。

```
public void mainperform(String inuptown, String inuptownid, int tt)
    {
        time++;
        Container panelin = getContentPane();
        gridbag = new GridBagLayout();
        panelin.setLayout(gridbag);
        uptownid = new String[100];
        uptownname = new String[100];
        uptownname[0] = new String(inuptown);
        getuptown();
        type = tt;
        int i = 0;
        do {
            i++;
        } while (!uptownname[i].equals(inuptown));
        buildingid = new String[150];
        getbuilding(inuptownid);
        buildingid[0] = new String("选择楼宇");
        roomid = new String[150];
        uptown = new String(inuptown);
        uptownid_select = new String(inuptownid);
        uptownlabel = new JLabel("选择小区");
```

```
uptownlabel.setToolTipText("点击选择小区");
inset = new Insets(5, 5, 5, 5);
c = new GridBagConstraints(2, 1, 1, 1, 0, 0, 10, 0, inset, 0, 0);
gridbag.setConstraints(uptownlabel, c);
panelin.add(uptownlabel);
```

在上面的代码中,通过 getContentPane 函数进行界面初始化操作。因为同一窗体根据不同的参数显示不同的功能,所以这种 GridBagLayout 网格布局的方式更便于查看。另外,通过 do...while 循环读取了符合条件的数据并展示出来。

(2) 用户可以查询不同的项目,在窗体的下拉列表框中进行选择。其代码如下。

```
uptownselect = new JComboBox(uptownname);
uptownselect.setSelectedIndex(i);
uptownselect.setMaximumRowCount(5);
uptownselect.addItemListener(new ItemListener() {
    public void itemStateChanged(ItemEvent event) {
        int i = 0;
        i = uptownselect.getSelectedIndex();
        t++;
        getbuilding(uptownid[i]);
        if (t < 2) {
            MonthDataInput mdif2 = new MonthDataInput(
                uptownname[i], uptownid[i], type);
        }
        MonthDataInput.this.setVisible(false);
    }
});
c = new GridBagConstraints(4, 1, 2, 1, 0, 0, 10, 0, inset, 0, 0);
gridbag.setConstraints(uptownselect, c);
panelin.add(uptownselect);
buildinglabel = new JLabel("选择楼宇");
buildinglabel.setToolTipText("点击选择楼宇");
c = new GridBagConstraints(2, 2, 1, 1, 0, 0, 10, 0, inset, 0, 0);
gridbag.setConstraints(buildinglabel, c);
panelin.add(buildinglabel);
buildingselect = new JComboBox(buildingid);
buildingselect.setMaximumRowCount(5);
buildingselect.addItemListener(new ItemListener() {
    public void itemStateChanged(ItemEvent event) {
        int i = 0;
        i = buildingselect.getSelectedIndex();
        inputbuilding = buildingid[i];
    }
});
c = new GridBagConstraints(4, 2, 2, 1, 0, 0, 10, 0, inset, 0, 0);
gridbag.setConstraints(buildingselect, c);
panelin.add(buildingselect);
```

(3) 当界面布局完毕后,为"确定"按钮增加监听器,执行数据库的访问操作。在访问数据库时,要根据查询类型执行不同条件的查询,其中,type 值用于区别不同的查询项目。其代码如下。

```java
button1 = new JButton("确定");
button1.addActionListener(new ActionListener() {
    public void actionPerformed(ActionEvent event) {
        if (type == 1) {
            MonthDataInput mdif3 = new MonthDataInput(uptown,
                    inputbuilding, uptownid_select);
            MonthDataInput.this.setVisible(false);
        } else if (type == 2) {
            ChargeReport elecReport = new ChargeReport("电费收费报表",
                    Integer.parseInt(uptownid_select), Integer
                            .parseInt(inputbuilding));
        } else if (type == 3) {
            ChargeReport waterReport = new ChargeReport("水费收费报表",
                    Integer.parseInt(uptownid_select), Integer
                            .parseInt(inputbuilding));
        } else if (type == 4) {
            ChargeReport gasReport = new ChargeReport("天燃气费收费报表",
                    Integer.parseInt(uptownid_select), Integer
                            .parseInt(inputbuilding));
        }
        MonthDataInput.this.setVisible(false);
    }
});
c = new GridBagConstraints(3, 4, 1, 2, 0, 0, 10, 0, inset, 0, 0);
gridbag.setConstraints(button1, c);
panelin.add(button1);
```

(4) 当用户查询数据完毕后,单击"返回"按钮关闭费用查询窗体,显示系统主界面。其代码如下。

```java
button2 = new JButton("返回");
button2.addActionListener(new ActionListener() {
    public void actionPerformed(ActionEvent event) {
        MonthDataInput.this.setVisible(false);
    }
});
c = new GridBagConstraints(5, 4, 1, 2, 0, 0, 10, 0, inset, 0, 0);
gridbag.setConstraints(button2, c);
panelin.add(button2);
panelin.repaint();
```

```
        panelin.setVisible(true);
        setSize(350, 300);
        setVisible(true);
    }
```

4.8 系统测试

项目运行后的主界面效果如图 4-16 所示。

扫码看视频

图 4-16　系统运行主界面

第 5 章 仿《羊了个羊》游戏

《羊了个羊》是一款由北京简游科技有限公司开发的休闲类益智游戏,于 2022 年 6 月 13 日正式发行。该游戏主要是让玩家利用各种道具和提示来消除每一个关卡当中的障碍和陷阱。游戏第一关其实是玩法教程,当玩家来到第二关时,难度直线上升。本章将介绍使用 Java 语言开发一个仿《羊了个羊》游戏的方法,并详细介绍整个项目的具体实现流程。本章项目由 IntelliJ+Swing 实现。

5.1 背景介绍

《羊了个羊》是曾经很火的一款微信游戏,在仿这款游戏之前,首先讲解一下游戏行业的背景,了解游戏行业的现状和发展前景。

扫码看视频

5.1.1 游戏行业发展现状

近年来,随着互联网、移动互联网技术的兴起和快速发展,互联网基础设施越来越完善,互联网用户规模迅速增长。受益于整个互联网产业的爆炸式增长,我国网络游戏产业呈现出飞速发展的态势,网络游戏整体用户规模持续扩大。《2022 年中国游戏产业报告》显示:2021 年中国游戏市场实际销售收入 2658.84 亿元,同比下降 10.33%;游戏用户规模达 6.64 亿,同比下降 0.33%;2022 年自主研发游戏国内市场实际销售收入 2223.77 亿元,同比下降 13.07%。

中国客户端游戏市场已步入成熟期,进入存量竞争阶段,面临来自移动游戏的竞争压力,行业内部竞争激烈,发展速度逐渐放缓。和客户端游戏一样,网页游戏也面临着移动游戏的竞争,并且日渐没落。自 2015 年起,中国网页游戏市场实际收入持续下降,2019 年收入仅为 98.7 亿元,比 2018 年减少 27.8 亿元,下降 22.0%。中国网页游戏市场近几年由于受到移动端市场的冲击,从事网页游戏的企业越来越少,用户逐步向移动游戏转移,人数由 2015 年的 3 亿人下降至 2022 年的 1.7 亿人。

5.1.2 虚拟现实快速发展

根据 Newzoo(权威数据机构)的 2022 年 VR(Virtual Reality,虚拟现实) 游戏市场报告的数据,到 2022 年年底,VR 游戏市场的当年收入达 18 亿美元。在未来几年里,这一市场的复合增长率将会达到 45%。Newzoo 还表示,随着游戏设备性能的提升,VR 用户正以前所未有的速度增长,至 2022 年年底达到 277 亿用户,预计到 2024 年年底将达到 460 亿。其中,独立 VR 耳机的增加是用户增长的主要原因之一,如无须额外硬件即可设置和使用 Meta Quest 设备,最近和即将推出的新产品也可能对 2026 年的收益产生影响。

电子游戏行业在对 VR 技术的掌握上已经投入了许多时间和资源,创造能投入市场的产品的尝试可以追溯到 20 世纪 80 年代,但在接口设备方面的进度一直停滞不前。2013 年,Oculus Rift 头戴显示器的推出带来了接口上的突破性进展,重新激发了游戏开发商和技术提供商对 VR 技术的兴趣。

而 2020 年 Valve《半条命:爱莉克斯》的到来,则全面解决了玩家的动作、数字环境中的物理存在感、与游戏内物体和环境的交互、可玩性以及通过 VR 媒体对故事进行叙述等更多 VR 游戏的问题，展示了商业产品的可行性，让越来越多的 PC 游戏开发商投身 VR 技术研发。

如今，VR 技术仍在不断发展，并吸引了大公司的投资，其中最为瞩目的当数 Meta Connect 2022 大会上，Meta 宣布收购了三家拥有深厚经验和成熟技术的 VR 工作室。在陆续收购了多家游戏工作室后，Meta 继续为其元宇宙事业招兵买马。

5.1.3 云游戏持续增长

2022 年，云游戏技术和服务取得了重大进展，如罗技和腾讯合作开发的手持游戏机罗技 G Cloud、雷蛇旗下支持 5G 连接的全新云驱动手持游戏设备 Razer Edge 5G 等。目前云游戏还处于发展阶段，从市场规模来看，被视为"游戏的未来"的云游戏行业正处于增长趋势中。

Newzoo 的 2022 年全球云游戏报告显示，在 2022 年，云游戏服务吸引了超过 3000 万名付费用户，这些用户的总消费额达 24 亿美元。到 2025 年，全球云游戏市场的年收入有望增长至 82 亿美元。Newzoo 表示，随着新的云游戏服务不断推出，云游戏市场正变得越来越成熟。在付费设计和使用场景等方面的创新，使得云游戏吸引了更多的用户并且能更好地满足其需求。与此同时，微软 Xbox 和索尼 PlayStation 等传统游戏巨头也在积极拓展云游戏业务。

美国软件公司 Perforce 在采访了 300 多名行业专业人士后，也得出结论：流媒体和云游戏将在不久的将来成为电子游戏的主要方式。随着 5G 覆盖范围的增加，甚至 6G 技术的发展，云游戏将进一步巩固其在游戏领域的地位。

5.1.4 移动游戏重回增长轨迹

Newzoo 表示，随着生活回归正常，移动游戏尤其是超休闲类游戏将迎来下滑，2022 移动业务收入下降 6.4%，至 922 亿美元。从长期来看，全球游戏市场将恢复增长轨迹，Newzoo 预计到 2025 年，游戏市场将创造 2112 亿美元的收益，年复合增长率为 3.4%，游戏市场未来几年很有前景，尤其是主机游戏。Sensor Tower 预计，全球手游市场营收也将在 2023 年重回增长轨迹，到 2026 年将上升至 1170 亿美元，未来几年内的年均复合增长率约为 5.6%。混合休闲等新趋势的兴起和中度手游的复苏，也将为手游市场带来新的活力。

近年来大厂纷纷进军手游市场，在备受关注的微软暴雪收购案中，Xbox 表示希望通过这一交易扩大其在移动游戏中的影响力；早些时候，微软还提到了打造 Xbox 手机游戏商店的计划。移动游戏已经越来越成为全球游戏市场的主导力量。

5.2 项目分析

在 2022 年，一款名为《羊了个羊》的消除小游戏突然爆火，全网热度居高不下，登上微博热搜榜第一。本节将简要介绍这款游戏的规则，规划整个项目的开发流程。

扫码看视频

5.2.1 游戏介绍

《羊了个羊》的整体设置非常简单，玩家只要在多层堆叠的方块中找到三个图案一致的方块并放入下方卡槽即可消除，方块全部消除便视为通关。和很多休闲类小游戏不同的是，《羊了个羊》只有两关，通过第二关后当天便不能继续游戏。并且，该游戏第一关难度极低，第二关难度突然飙升。用网友的话说："第一关瞎玩都能过，第二关玩瞎了都过不去。"

《羊了个羊》的游戏规则和特点如下。

- ❑ 玩家点击屏幕中间亮起的方块，并将其移动到下面的卡槽中。如果三个方块的图案相同，就会被消除。如果卡槽被填满，游戏就会失败。
- ❑ 玩家移动方块时，只能移动翻牌区的方块，也就是最上面的方块。下面的方块必须在上面的方块移除后变亮，才能移动。
- ❑ 游戏中可以使用辅助道具帮助通关，但是道具的使用次数是有限的。

通过上述描述可以看出，《羊了个羊》跟连连看、多层连连看、爱消除等休闲类小游戏类似，所以在《羊了个羊》游戏爆火之后，抄袭的话题也很快登上了微博热搜，其被质疑抄袭同类游戏《3tiles》，因为两者在画面、玩法上确实非常相似。不过，抛开是否抄袭暂且不谈，《羊了个羊》能够在短时间内爆火，原因并非其玩法，而在于其背后的竞技游戏思维。

5.2.2 规划开发流程

要想做好《羊了个羊》游戏项目的功能分析，需要将这款游戏从头到尾试玩几次，彻底了解《羊了个羊》游戏的具体玩法和过程，然后根据游戏规则和要求总结出游戏的基本功能模块。

根据软件项目的开发流程，可以做一个简单的项目规划书。整个规划书分为以下两个部分。

- ❑ 系统需求分析。
- ❑ 结构规划。

《羊了个羊》游戏项目的开发流程如图 5-1 所示。

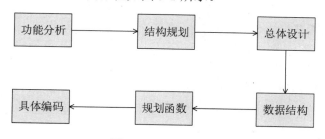

图 5-1　开发流程图

- 功能分析：分析整个系统所需要的功能。
- 结构规划：规划系统中所需要的功能模块。
- 总体设计：分析系统处理流程，探索系统核心模块的运作。
- 数据结构：设计系统中需要的数据结构。
- 规划函数：规划系统中需要的功能函数。
- 具体编码：编写系统的具体实现代码。

5.2.3　模块结构

在结构规划阶段，为了加深印象，可以绘制一个模块结构图，如图 5-2 所示。

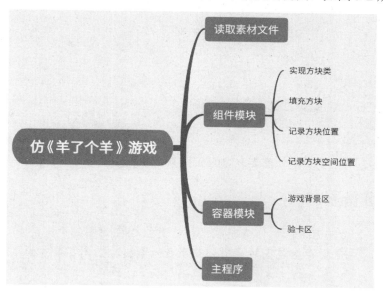

图 5-2　游戏模块结构

5.3 准备工作

根据上面的需求分析和游戏模块结构,开始制订整个开发计划。在开始具体编码工作之前,先完成必需的准备工作。

扫码看视频

5.3.1 创建工程

本项目采用 IntelliJ IDEA 工具实现,首先使用 IntelliJ IDEA 创建一个名为 TestOfSheep 的工程,如图 5-3 所示。

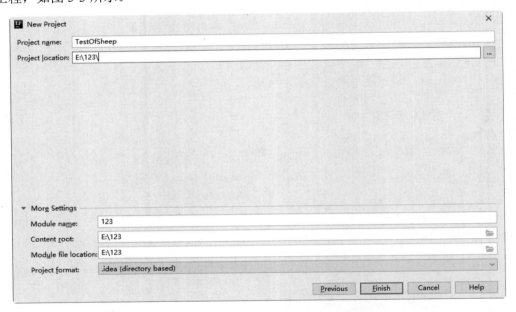

图 5-3 创建工程 TestOfSheep

5.3.2 准备素材

本项目需要大量的方块图片作为游戏素材,并且还需要用到不同的背景音乐素材。创建 static 文件夹,用于保存上述图片素材和背景音乐素材,如图 5-4 所示。

图 5-4 工程 TestOfSheep 的目录结构

5.4 读取素材文件

在项目的方块中需要显示预先准备好的素材图片，方块中显示的图片是随机的。因为在系统中需要多次操作图片并使用背景音乐，所以单独创建一个文件夹来保存素材文件的源码。创建 util 文件夹，在里面编写文件 ReadResourceUtil.java，功能是读取 static 文件夹中的图片素材和背景音乐素材。文件 ReadResourceUtil.java 的具体实现代码如下。

扫码看视频

```
/**
 * 外部资源加载类，可以从 jar 中读取内容
 */
public class ReadResourceUtil {
    static Random random = new Random();
    static File jarFile;
    static {
        try {
            jarFile = new File(java.net.URLDecoder.decode(ReadResourceUtil.class.
        getProtectionDomain().getCodeSource().getLocation().getFile(), "UTF-8"));
        } catch (UnsupportedEncodingException e) {
            e.printStackTrace();
        }
    }
    /**
     * 读取音频
     */
```

```java
    public static URL readAudio(String name){
        return getUri("/static/audio/wav" +random.nextInt(2)+"/"+name);
    }
    public static URL getUri(String name){
        URL resource = ReadResourceUtil.class.getResource(name);
        if(resource==null){
            try {
                File file = new File(jarFile.getParentFile(), "static");
                System.out.println(file.getAbsolutePath()+"\t"+file.exists());
                if(file.exists()) {
                    resource = new File(jarFile.getParentFile(), name).toURL();
                }else{
                    resource = new File(jarFile.getParentFile().getParentFile(),
                            name).toURL();
                }
            } catch (MalformedURLException e) {
                e.printStackTrace();
            }
        }
        return resource;
    }
    /**
     * 读取图片
     */
    private static String readSkinPath(boolean isRandom){
        if(isRandom){
            return "/static/skins/a" +random.nextInt(2);
        }else{
            return "/static/skins/a3";
        }
    }
    /**
     * 读取图片
     */
    public static List<String> readSkin(boolean isRandom){
        String path = readSkinPath(isRandom);
        System.out.println(path);
        List<String> list = new ArrayList<>();
        try {
            URL url = getUri("/" + path);
            if (url != null) {
                try {
                    final File apps = new File(url.toURI());
                    for (File app : apps.listFiles()) {
                        list.add(path+"/"+app.getName());
                    }
                } catch (URISyntaxException ex) {
```

```
                ex.printStackTrace();
            }
        }
    }catch (Exception e){
        e.printStackTrace();
    }
    return list;
    }
}
```

5.5 组件模块

整个游戏界面由多个方块组成,每个方块内放置了随机图片,游戏玩家在游戏过程中可以通过鼠标单击或拖曳方块。本游戏的核心是方块和方块的拖曳操作,本项目通过组件模块实现这一核心功能,由以下两个文件夹中的程序文件实现。

扫码看视频

- 文件夹 drager:实现方块类。
- 文件夹 components:将图片素材类添加到方块中。

5.5.1 实现方块类

在文件夹 drager 中编写文件 ShapeDrager.java,设置方块被拖曳时的样式,并且获取方块的旋转角度。文件 ShapeDrager.java 的具体实现代码如下。

```java
public abstract class ShapeDrager extends JPanel {
    // 旋转的角度
    double rotation = 0;
    // 是否为图标模式
    boolean iconType = false;
    public ShapeDrager(JComponent parent){
        this(parent,false);
    }
    public ShapeDrager(JComponent parent, boolean iconType){
        this.iconType = iconType;
        this.setFocusable(true);
        if(!iconType) {
            this.setCursor(new Cursor(Cursor.MOVE_CURSOR));
        }else{
            this.setCursor(new Cursor(Cursor.HAND_CURSOR));
        }
        initShape();
```

```
    }
    private void initShape(){
        this.setOpaque(false);
        this.setLayout(new BorderLayout());
    }
    public double getRotation() {
        return rotation;
    }
}
```

5.5.2 填充方块

接下来需要在方块中添加图片素材，即在文件夹 components 中创建填充方块的文件。编写文件 Fruits.java，功能是向方块中填充随机素材图片。文件 Fruits.java 的具体实现代码如下。

```
/**
 * 方块对象
 */
public class Fruits extends ShapeDrager {

    // 方块上放置的图形
    Image image = null;
    // 方块名称
    String imageName;

    // 方块圆角边框
    int arc = 10;
    // 透明度
    float alpha = 1;
    Color bgColor = new Color(138,158,58);
    public Fruits(JPanel parent,Image image,String imageName){
        super(parent);
        this.image = image;
        this.imageName = imageName;
    }

    @Override
    protected void paintComponent(Graphics g) {
        Graphics2D g2d = (Graphics2D) g;
        g2d.setRenderingHint(RenderingHints.KEY_ANTIALIASING,
                             RenderingHints.VALUE_ANTIALIAS_ON);
        BasicStroke basicStroke = new BasicStroke(2);
        g2d.setStroke(basicStroke);
        if(alpha<1) {
```

```
            g2d.setComposite(AlphaComposite.getInstance(AlphaComposite.SRC_OVER,
                            alpha));
        }
        g2d.rotate(getRotation(),getWidth()/2,getHeight()/2);
        // 绘制底色
        g2d.setColor(bgColor);
        g2d.fillRoundRect(0, 0, (int) (getSize().width - 1), (int) (getSize().height
                          - 1),arc,arc);
        // 绘制白色
        g2d.setColor(Color.WHITE);
        g2d.fillRoundRect(0, 0, (int) (getSize().width - 1), (int) (getSize().height
                          - 1-5),arc,arc);
        // 绘制外边框
        g2d.setColor(bgColor);
        g2d.drawRoundRect(0, 0, (int) (getSize().width - 1), (int) (getSize().height
                          - 1),arc,arc);
        // 绘制图片
        if(image!=null) {
            int imageWidth = image.getWidth(null);
            int imageHeight = image.getHeight(null);
            int x = (getWidth()-imageWidth)/2;
            int y = (getHeight()-5-imageHeight)/2;
            g2d.drawImage(image,x,y,imageWidth,imageHeight,null);
        }
        super.paintComponent(g);
    }
    public String getImageName() {
        return imageName;
    }
    public void setAlpha(float alpha) {
        this.alpha = alpha;
        repaint();
    }
}
```

5.5.3 记录方块位置

编写文件 FruitObject.java，功能是创建方块对象并放置位置对象，同时记录方块被放置的位置。文件 FruitObject.java 的具体实现流程如下。

(1) 创建类 FruitObject，分别设置方块的宽度和高度，并用自定义方法获取每一个方块所在的层和坐标。对应代码如下。

```
public class FruitObject {
    // 方块默认宽度
    public static final int defaultWidht = 100;
```

```java
// 方块默认高度
public static final int defaultHeight = 100;
static Random RANDOM = new Random();
// 点击方块的声音
static AudioClip audioClip;
static {
    try {
        audioClip = AudioClip.load(ReadResourceUtil.readAudio("click.wav").toURI());
    } catch (Exception e) {
        e.printStackTrace();
    }
}
// 存放的方块
Fruits fruits;
// 所在层的x序号
int x;
// 所在层的y序号
int y;
// 所在层序号
int level;
String imageName;                       //名称,非常重要,按照名称来判断是否为同一类型
boolean flag = true;                    //是否可以点击
boolean leftFold = false;               //是否为左折叠元素
boolean rightFold = false;              //是否为右折叠元素
ImageCantainer imageCantainer;
CardSlotCantainer cardSlotCantainer;

public FruitObject(CardSlotCantainer cardSlotCantainer,Fruits fruits, int x,
                int y, int level) {
    this.fruits = fruits;
    this.x = x;
    this.y = y;
    this.level = level;
    this.imageName = fruits.getImageName();
    this.cardSlotCantainer = cardSlotCantainer;
}

public Fruits getFruits() {
    return fruits;
}

public void setX(int x) {
    this.x = x;
}

public void setY(int y) {
    this.y = y;
```

```
    }

    public int getLevel() {
        return level;
    }

    public void setLevel(int level) {
        this.level = level;
    }

    public String getImageName() {
        return imageName;
    }
```

(2) 创建方法 show()，功能是列表展示各个方块，将不同的素材图片放到方块中，并将方块放在不同的层中，形成错落有致的效果。对应代码如下。

```
/**
 * 添加到叠卡区
 */
public void show(ImageCantainer imageCantainer, int initX, int initY) {
    this.imageCantainer = imageCantainer;
    // 随机生成坐标偏移量，实现上下层错落有致的视觉感
    boolean ranDomWidth = RANDOM.nextInt(10)%2==0;
    boolean ranDomHeight = RANDOM.nextInt(10)%2==0;
    // 通过在 x 和 y 方向上随机选择是否偏移，默认宽度/2 或默认高度/2
    int pointX = initX + x*defaultWidht+(ranDomWidth?defaultWidht/2:0);
    int pointY = initY + y*defaultHeight+(ranDomHeight?defaultHeight/2:0);
    // 设置方块在背景面板中的位置
    fruits.setBounds(pointX,pointY,defaultWidht,defaultHeight);
    // 记录方块的空间信息
    SpaceManager.rectangle(this);
    imageCantainer.add(fruits,0);
    addClick();
}
```

(3) 创建方法 showFold()，功能是将指定的方块添加到翻牌区。对应代码如下。

```
/**
 * 添加到翻牌区
 * @param imageCantainer
 */
public void showFold(ImageCantainer imageCantainer, int initX, int initY,
                    int offset, boolean isLeft) {
    this.imageCantainer = imageCantainer;
    // 随机生成开始坐标偏移量，实现上下层错落有致的视觉感
    int pointX = initX + x*defaultWidht+offset;
    int pointY = initY + y*defaultHeight-defaultHeight/4;
```

```
        if(isLeft){
            this.leftFold = true;
        }else{
            this.rightFold= true;
        }
        // 设置方块在背景面板中的位置
        fruits.setBounds(pointX,pointY,defaultWidht,defaultHeight);
        // 记录方块的空间信息
        SpaceManager.rectangle(this);
        imageCantainer.add(fruits,0);
        addClick();
    }
```

(4) 创建方法 addClick(),功能是监听用户点击鼠标事件,被点击的方块会添加到卡槽。对应代码如下。

```
    /**
     * 方块点击事件:点击后添加到卡槽
     */
    public void addClick(){
        fruits.addMouseListener(new MouseAdapter() {
            @Override
            public void mouseClicked(MouseEvent e) {
                if(flag) {
                    if (audioClip != null) {
                        audioClip.play();
                    }
                    cardSlotCantainer.addSlot(FruitObject.this);
                }
            }
        };
    }
```

(5) 编写方法 removeImageCantainer(),移除符合规则的方块;编写方法 removeCardSlotCantainer(),移除卡槽容器;编写方法 setFlag(boolean flag),设置指定的图片不可被移动。对应代码如下。

```
    public void removeImageCantainer() {
        Rectangle visibleRect = fruits.getVisibleRect();
        SpaceManager.removeCompontFlag(this);
        imageCantainer.remove(fruits);
        imageCantainer.repaint(visibleRect);
    }

    public void removeCardSlotCantainer() {
        cardSlotCantainer.remove(fruits);
    }
```

```java
public boolean isLeftFold() {
   return leftFold;
}
/**
 * 设置
 */
public void setFlag(boolean flag) {
   if(this.flag!=flag){                    //需要更新透明度
      getFruits().setAlpha(flag?1:0.2f);
   }
   this.flag = flag;
}
```

5.5.4 记录方块空间位置

编写文件 SpaceManager.java，功能是记录每层方块的空间信息，用于检测上下层是否有符合规则的方块，验证是否可以通过鼠标点击实现动态刷新功能。文件 SpaceManager.java 的具体实现流程如下。

（1）编写方法 rectangle()，记录方块坐标重叠数据，实现代码如下。

```java
public class SpaceManager {
   // Map<X 坐标点, Set<层数, HashMap<Y 坐标点 Integer, FruitObject>>>
   static Map<Integer, Map<Integer, List<FruitObject>>> LEAVL_DATA_X1 = new
                  HashMap<>();
   static Map<Integer, Map<Integer, List<FruitObject>>> LEAVL_DATA_X2 = new
                  HashMap<>();
   static Map<Integer, Map<Integer, List<FruitObject>>> LEAVL_DATA_X3 = new
                  HashMap<>();
   static Map<Integer, Map<Integer, List<FruitObject>>> LEAVL_DATA_Y1 = new
                  HashMap<>();
   static Map<Integer, Map<Integer, List<FruitObject>>> LEAVL_DATA_Y2 = new
                  HashMap<>();
   static Map<Integer, Map<Integer, List<FruitObject>>> LEAVL_DATA_Y3 = new
                  HashMap<>();

   public static void rectangle(FruitObject fruitObject){
      Rectangle bounds = fruitObject.fruits.getBounds();
      if(fruitObject.isLeftFold()){
         bounds.x=0;bounds.y=0;
      }
      if(fruitObject.isRightFold()){
         bounds.x=1;bounds.y=1;
      }
```

```
            int x1 = bounds.x;
            int x2 = bounds.x+FruitObject.defaultWidht/2;
            int x3 = bounds.x+FruitObject.defaultHeight;
            int y1 = bounds.y;
            int y2 = bounds.y+FruitObject.defaultHeight/2;
            int y3 = bounds.y+FruitObject.defaultHeight;
            putToCache(x1,fruitObject,LEAVL_DATA_X1);
            putToCache(x2,fruitObject,LEAVL_DATA_X2);
            putToCache(x3,fruitObject,LEAVL_DATA_X3);
            putToCache(y1,fruitObject,LEAVL_DATA_Y1);
            putToCache(y2,fruitObject,LEAVL_DATA_Y2);
            putToCache(y3,fruitObject,LEAVL_DATA_Y3);
            updateCompontFlag(fruitObject,false,bounds);
        }
```

(2) 编写方法 putToCache()，将指定的方块放入缓存中，代码如下。

```
    private static void putToCache(int point,FruitObject fruitObject,Map<Integer,
                                Map<Integer, List<FruitObject>>> LEAVL_DATA){
        Map<Integer, List<FruitObject>> hashLevel = LEAVL_DATA.get(fruitObject.level);
        if(hashLevel==null){
            hashLevel = new HashMap<>();
        }
        List<FruitObject> objects = hashLevel.get(point);
        if(objects==null){
            objects = new ArrayList<>();
        }
        objects.add(fruitObject);
        hashLevel.put(point,objects);
        LEAVL_DATA.put(fruitObject.level,hashLevel);
    }
```

(3) 编写方法 removeCompontFlag()，功能是当消除方块时计算下一层被点亮的方块，代码如下。

```
    public static void removeCompontFlag(FruitObject fruitObject){
        Rectangle bounds = fruitObject.fruits.getBounds();
        if(fruitObject.isLeftFold()){
            bounds.x=0;bounds.y=0;
        }
        if(fruitObject.isRightFold()){
            bounds.x=1;bounds.y=1;
        }
        int x1 = bounds.x;
        int x2 = bounds.x+FruitObject.defaultWidht/2;
        int x3 = bounds.x+FruitObject.defaultHeight;
        int y1 = bounds.y;
        int y2 = bounds.y+FruitObject.defaultHeight/2;
```

```
        int y3 = bounds.y+FruitObject.defaultHeight;
        deleteLevelFlag(x1,fruitObject,LEAVL_DATA_X1);
        deleteLevelFlag(x2,fruitObject,LEAVL_DATA_X2);
        deleteLevelFlag(x3,fruitObject,LEAVL_DATA_X3);
        deleteLevelFlag(y1,fruitObject,LEAVL_DATA_Y1);
        deleteLevelFlag(y2,fruitObject,LEAVL_DATA_Y2);
        deleteLevelFlag(y3,fruitObject,LEAVL_DATA_Y3);
        updateCompontFlag(fruitObject,true,bounds);
    }
```

(4) 编写方法 deleteLevelFlag()，功能是检测当前是否有符合规则的方块，并消除这些方块，对应代码如下。

```
    private static void deleteLevelFlag(int point,FruitObject 
fruitObject,Map<Integer, Map<Integer, List<FruitObject>>> LEAVL_DATA){
        Map<Integer, List<FruitObject>> hashLevel = LEAVL_DATA.get(fruitObject.getLevel());
        if(hashLevel==null){
            return;
        }
        List<FruitObject> objects = hashLevel.get(point);//删除对应层对应点的坐标
        if(objects==null){
            hashLevel.remove(point);
            if(hashLevel.size()==0){
                LEAVL_DATA.remove(fruitObject.getLevel());
            }else {
                LEAVL_DATA.put(fruitObject.getLevel(), hashLevel);
            }
            return;
        }
        objects.remove(fruitObject);//从当前层次和点的坐标列表中移除当前方块
        if(objects.isEmpty()){
            hashLevel.remove(point);
            if(hashLevel.size()==0){
                LEAVL_DATA.remove(fruitObject.getLevel());
            }else {
                LEAVL_DATA.put(fruitObject.getLevel(), hashLevel);
            }
        }
    }
```

(5) 编写方法 updateCompontFlag()，功能是根据给定的坐标和层次更新图层节点的状态，false 表示修改为不可点击，反之表示可点击。对应代码如下。

```
    private static void updateCompontFlag(FruitObject fruitObject,boolean 
flag,Rectangle bounds){
        int level = fruitObject.getLevel();
        int x1 = bounds.x;
```

```
        int x2 = bounds.x+FruitObject.defaultWidht/2;
        int x3 = bounds.x+FruitObject.defaultHeight;
        int y1 = bounds.y;
        int y2 = bounds.y+FruitObject.defaultHeight/2;
        int y3 = bounds.y+FruitObject.defaultHeight;
        for (int i = 0; i<level; i++) {
            updateLevelFlag(x1,x2,x3,y1,y2,y3,x1,i,LEAVL_DATA_X1,flag);
            updateLevelFlag(x1,x2,x3,y1,y2,y3,x2,i,LEAVL_DATA_X2,flag);
            updateLevelFlag(x1,x2,x3,y1,y2,y3,x3,i,LEAVL_DATA_X3,flag);
            updateLevelFlag(x1,x2,x3,y1,y2,y3,y1,i,LEAVL_DATA_Y1,flag);
            updateLevelFlag(x1,x2,x3,y1,y2,y3,y2,i,LEAVL_DATA_Y2,flag);
            updateLevelFlag(x1,x2,x3,y1,y2,y3,y3,i,LEAVL_DATA_Y3,flag);
        }
    }
```

(6) 编写方法 updateLeveFlag()，功能是删除符合规则的方块后更新当前层的显示信息。对应代码如下。

```
    private static void updateLevelFlag(int x1,int x2,int x3,int y1,int y2,int y3,int point,int level,Map<Integer, Map<Integer, List<FruitObject>>> LEAVL_DATA,boolean flag){
        Map<Integer, List<FruitObject>> hashLevel = LEAVL_DATA.get(level);
        if(hashLevel==null){
            return;
        }
        List<FruitObject> objects = hashLevel.get(point);
        if(objects==null){
            return;
        }
        if(flag) {             // 当更新的时候需要强制检查
            flag = !LEAVL_DATA.containsKey(level + 1);    //上一层都没了，可以点击
        }
        //上一层存在，但是没有遮挡，可以点击
        if(!flag){
            flag = !(isExsit(x1,level+1,LEAVL_DATA_X1)
                    && isExsit(x2,level+1,LEAVL_DATA_X2)
                    && isExsit(x3,level+1,LEAVL_DATA_X3)
                    && isExsit(y1,level+1,LEAVL_DATA_Y1)
                    && isExsit(y2,level+1,LEAVL_DATA_Y2)
                    && isExsit(y3,level+1,LEAVL_DATA_Y3));
        }
        for (FruitObject object : objects) {
            object.setFlag(flag);
        }
    }

    private static boolean isExsit(int point,int level,Map<Integer, Map<Integer, List<FruitObject>>> LEAVL_DATA){
```

```
            Map<Integer, List<FruitObject>> hashLevel = LEAVL_DATA.get(level);
            if(hashLevel==null){
                return false;
            }
            List<FruitObject> objects = hashLevel.get(point);
            if(objects==null||objects.isEmpty()){
                return false;
            }
            return true;
    }
}
```

5.6 容器模块

创建 cantainer 文件夹实现容器模块,制作整个游戏界面的背景,并在卡槽验证用户添加的方块是否符合游戏规则。

扫码看视频

5.6.1 游戏背景区

编写文件 ImageCantainer.java 实现游戏背景区,绘制整个游戏背景,并能够根据用户的操作,实现游戏背景的放大和缩小功能。具体实现代码如下。

```
public class ImageCantainer extends JPanel implements ScaleFunction {
    // 背景颜色
    Color backgroudColor = Color.WHITE;
    // 间距
    int step = 20;
    // 缩放比例,每次缩放一格增减 5
    int scale = 100;
    public ImageCantainer(){
        this.setLayout(null);
        this.setBorder(new EmptyBorder(20,20,20,20));
        this.setBackground(Color.RED);
        this.setCursor(new Cursor(Cursor.CROSSHAIR_CURSOR));
    }
    // 背景色
    Color bgColor = new Color(195,254,139);
    // 背景图标色
    Color grassColor = new Color(95,154,39);
    Random random =  new Random();
    
    public void drawX(Graphics g){
        int size = 20;
```

```java
        int width = (int) (this.getWidth()/(scale/100F));
        int height = (int) (this.getHeight()/(scale/100F));
        int skip = (int) (step*scale/100F);
        Graphics2D graphics = (Graphics2D)g;
        graphics.setRenderingHint(RenderingHints.KEY_ANTIALIASING,
                        RenderingHints.VALUE_ANTIALIAS_ON);
        // 背景色
        graphics.setColor(bgColor);
        graphics.fillRect(0,0,width,height);
        // 设置画笔颜色或大小
        graphics.setColor(grassColor);
        BasicStroke basicStroke = new BasicStroke(2);
        graphics.setStroke(basicStroke);
        // 随机绘制各种大小图形
        for (int i = 0; i < width; i+=skip) {
            graphics.drawRect(random.nextInt(width),random.nextInt(height),
                        random.nextInt(size),random.nextInt(size));
            graphics.drawArc(random.nextInt(width),random.nextInt(height),
random.nextInt(size),random.nextInt(size),random.nextInt(360),random.nextInt(360));
            graphics.drawOval(random.nextInt(width),random.nextInt(height),
                        random.nextInt(size),random.nextInt(size));
            graphics.drawRoundRect(random.nextInt(width),random.nextInt(height),
random.nextInt(size),random.nextInt(size),random.nextInt(360),random.nextInt(360));
        }
    }

    @Override
    protected void paintComponent(Graphics graphics) {
        Graphics2D g = (Graphics2D)graphics;
        clearPane(g);
        AffineTransform transform = new AffineTransform(g.getTransform());
        g.setTransform(transform);
        g.scale(scale/100F,scale/100F);
        graphics.setColor(backgroudColor);
        graphics.fillRect(0,0,getWidth(),getHeight());
        drawX(graphics);
        transform.concatenate(g.getTransform());
    }

    // 清空背景
    private void clearPane(Graphics2D g) {
        g.clearRect(0, 0, getWidth(), getHeight());
    }
    /**
     * 重置大小
     */
    @Override
```

```
public void reset(){
   scale=100;
   scale(0);
}
/**
 * 放大或者缩小
 * @param step
 */
@Override
public void scale(int step){
   scale+=step;
   if(scale<20){
      scale=20;
   }
   if(scale>500){
      scale=500;
   }
   applyZoom();
}
private void applyZoom() {
   setPreferredSize(new Dimension((int) (scale/100F * getWidth()), (int) (scale/100F * getHeight())));
   validate();
   repaint();
}
}
```

5.6.2 卡槽

编写文件 CardSlotCantainer.java 实现卡槽功能,验证用户添加的方块是否符合要求。具体实现代码如下。

(1) 创建卡槽类 CardSlotCantainer,设置卡槽中存放的方块最多是 7,在卡槽中绘制对应的方块。实现代码如下。

```
public class CardSlotCantainer extends JPanel {
   static AudioClip audioClip;
   static AudioClip failClip;
   static {
      try {
         audioClip = AudioClip.load(ReadResourceUtil.readAudio("win.wav").toURI());
      } catch (Exception e) {
         e.printStackTrace();
      }
      try {
```

```
            failClip = AudioClip.load(ReadResourceUtil.readAudio("fail.wav").toURI());
        } catch (Exception e) {
            e.printStackTrace();
        }
    }
    // 方块间隔
    int step = 5;
    // 内外底色的间隔
    int borderSize = 10;
    // 方块最大数
    int solt = 7;
    // 是否结束
    boolean isOver = false;
    Color bgColor = new Color(157,97,27);
    Color borderColor = new Color(198,128,48);
    // 圆角幅度
    int arc = 15;
    List<FruitObject> slot = new ArrayList<>();
    int initX;
    int initY;
    public CardSlotCantainer(ImageCantainer imageCantainer,int initX, int initY){
        initY+=20;
        initX+=-borderSize;
        this.setBorder(new EmptyBorder(0,0,0,0));
        this.setOpaque(false);
        this.setFocusable(true);
        this.setLayout(new FlowLayout());
        this.initY = initY;
        this.initX = initX;
        setBounds(initX,initY, FruitObject.defaultWidht*solt+step*2+borderSize*2, FruitObject.defaultHeight+step*2+borderSize*2);
        imageCantainer.add(this);
    }
```

(2) 编写方法 addSlot()，监听用户的鼠标事件，将被点击的方块添加到卡槽中进行验证。实现代码如下。

```
    public void addSlot(FruitObject object){
        if(isOver){
            return;
        }
        slot.add(object);
        // 卡槽的方块删除点击事件
        object.removeImageCantainer();
        MouseListener[] mouseListeners = object.getFruits().getMouseListeners();
        if(mouseListeners!=null){
            for (MouseListener mouseListener : mouseListeners) {
                object.getFruits().removeMouseListener(mouseListener);
```

```java
        }
    }
    // 排序卡槽中的图片
    slot.sort(Comparator.comparing(FruitObject::getImageName));
    // 3张相同图片的判断,如果有直接消除,思路是：分组后查看每组数量是否超过3张,如果超过则消除
    Map<String, List<FruitObject>> map = slot.stream().collect
            (Collectors.groupingBy(FruitObject::getImageName));
    Set<String> keys = map.keySet();
    for (String key : keys) {
        List<FruitObject> objects = map.get(key);
        if(objects.size()==3){
            if(audioClip!=null){
                audioClip.play();
            }
            // 消除的元素直接从集合中删除
            for (FruitObject fruitObject : objects) {
                fruitObject.removeCardSlotCantainer();
            }
            slot.removeAll(objects);
        }
    }
    // 新添加的方块,显示到卡槽中
    redraw();
    // 判断游戏是否结束
    if(slot.size()==solt){
        isOver = true;
        failClip.play();
        JOptionPane.showMessageDialog(this.getParent(), "Game Over：槽满了",
                "Tip",JOptionPane.ERROR_MESSAGE);
    }
}
```

(3) 当有新的方块被添加到卡槽时，需要及时更新游戏界面。更新功能通过以下代码实现。

```java
public void redraw(){
    this.removeAll();
    for (int i = 0; i < slot.size(); i++) {
        FruitObject fruitObject = slot.get(i);
        int pointX = step+i*FruitObject.defaultWidht+borderSize/2;
        fruitObject.getFruits().setBounds(pointX,borderSize,
                FruitObject.defaultWidht,FruitObject.defaultHeight+step);
        this.add(fruitObject.getFruits());
    }
    repaint();
}
```

```java
@Override
protected void paintComponent(Graphics g) {
    Graphics2D g2d = (Graphics2D) g;
    g2d.setRenderingHint(RenderingHints.KEY_ANTIALIASING,
                    RenderingHints.VALUE_ANTIALIAS_ON);
    BasicStroke basicStroke = new BasicStroke(borderSize);
    g2d.setStroke(basicStroke);
    // 绘制第 1 层底色
    g2d.setColor(bgColor);
    g2d.fillRoundRect(0, 0, (int) (getSize().width - borderSize), (int)
                    (getSize().height - borderSize),arc,arc);
    // 绘制第 2 层底色
    g2d.setColor(borderColor);
    g2d.fillRoundRect(borderSize, borderSize, (int) (getSize().width -
            1-borderSize*3), (int) (getSize().height - 1-borderSize*3),arc,arc);
    super.paintComponent(g);
    }
}
```

5.7 主程序

本项目的主程序文件是 UmlPanel.java，功能是调用前面介绍的模块实现整个游戏界面的初始化，将随机生成的方块用叠加的方式显示出来。为了便于调试，文件特意使用打印函数在控制台中输出和游戏相关的信息。文件 UmlPanel.java 的具体实现代码如下：

扫码看视频

```java
public class UmlPanel extends JPanel {
    ImageCantainer imageCantainer = new ImageCantainer();
    public UmlPanel(){
        this.setLayout(null);
        this.add(imageCantainer);
        this.addComponentListener(new ComponentAdapter() {
            @Override
            public void componentResized(ComponentEvent e) {
                reDrawImagePanel();
            }
        });
        initContent();
    }
    public void initContent(){
        int maxLevel = 10;      //多少层
        int maxWidth = 6;       //跨度个数
        int maxHeight = 5;      //最大宽度
        int maxFlop = 30;       //翻牌数量
```

```java
Random random = new Random();
// 如果需要随机皮肤，修改为true即可
List<String> list = ReadResourceUtil.readSkin(false);
int typeSize = list.size();// 多少种类
System.out.println("种类："+typeSize);
int groupNumber = (int) Math.ceil((maxLevel*maxWidth*maxHeight+maxFlop)/(3f
                * typeSize));// 求得每种种类的个数
System.out.println("每种组数："+groupNumber);
int groupCount = groupNumber*3;
System.out.println("每种总数："+groupCount);
System.out.println("共计数量："+(typeSize*groupCount+maxFlop));
// 绘制卡槽
int initX = 100;
int initY = 50;
CardSlotCantainer cardSlotCantainer = new
                CardSlotCantainer(imageCantainer,initX+((maxWidth-7)*
FruitObject.defaultWidht)/2,+initY+FruitObject.defaultHeight*(maxHeight+2));
// 随机生成方块集合：注意打乱顺序
List<FruitObject> objects = new ArrayList<>();
for (String temp : list) {
    try {
        BufferedImage bufferedImage =
                    ImageIO.read(ReadResourceUtil.getUri("/"+temp));
        int count = groupCount+(maxFlop>0?random.nextInt(maxFlop):0);
        for (int i = 0; i < count; i++) {
            int size = objects.size()-1;
            Fruits fruits = new Fruits(imageCantainer,bufferedImage,temp);
            fruits.setPreferredSize(new Dimension(100, 100));
            int index = 0;
            if(size>10) {
                index = random.nextInt(size);
            }
            objects.add(index, new FruitObject(cardSlotCantainer,fruits, 0, 0, 0));
            maxFlop--;
        }
    } catch (IOException e) {
        throw new RuntimeException(e);
    }
}
System.out.println("实际数量："+objects.size());
// 给每个对象设置坐标
int index = 0;
for (int i = 0; i < maxLevel; i++) {
    for (int x = 0; x < maxWidth; x++) {
        for (int y = 0; y < maxHeight; y++) {
            FruitObject fruitObject = objects.get(index++);
            fruitObject.setX(x);
```

```java
                    fruitObject.setY(y);
                    fruitObject.setLevel(i);
                    fruitObject.show(imageCantainer,initX-FruitObject.defaultWidht/4,
                                initY-FruitObject.defaultHeight/4);
            }
        }
    }
    System.out.println("重叠数量："+index);
    // 绘制翻牌区
    int size = objects.size();
    if(index<size){
        int step = 5;
        int lenght = (size-index)/2;
        for (int i = 0; i < lenght; i++) {     //绘制左边
            FruitObject fruitObject = objects.get(index++);
            fruitObject.setX(0);
            fruitObject.setY(maxHeight+1);
            fruitObject.setLevel(i);
            fruitObject.showFold(imageCantainer,initX+(lenght*step),
                            initY,-i*step,true);
        }
        System.out.println("左翻牌区数："+lenght+"\t"+index);
        lenght = size-index;
        for (int i = 0 ; i < lenght; i++) {   //绘制左边
            FruitObject fruitObject = objects.get(index++);
            fruitObject.setX(maxWidth-1);
            fruitObject.setY(maxHeight+1);
            fruitObject.setLevel(i);
            fruitObject.showFold(imageCantainer,initX-(lenght*step),
                            initY,i*step,false);
        }
        System.out.println("右翻牌区数："+lenght+"\t"+index);
    }
}

private void reDrawImagePanel(){
    int leftX = 15;
    int width = getWidth()-leftX-15;
    int height = getHeight()-leftX-15;
    imageCantainer.setBounds(leftX, leftX,width,height);
    imageCantainer.setPreferredSize(new Dimension(width, height));
    imageCantainer.setLayout(null);
}

public static void main(String[] args) {
    JFrame jFrame = new JFrame("yang 了 yang");
    jFrame.setLayout(new BorderLayout());
```

```
        jFrame.setPreferredSize(new Dimension(900,1000));
        jFrame.setDefaultCloseOperation(JFrame.EXIT_ON_CLOSE);
        jFrame.add(new UmlPanel(),BorderLayout.CENTER);
        jFrame.setVisible(true);
        jFrame.pack();
    }
}
```

5.8 调试运行

执行本项目后的界面效果如图 5-5 所示。

扫码看视频

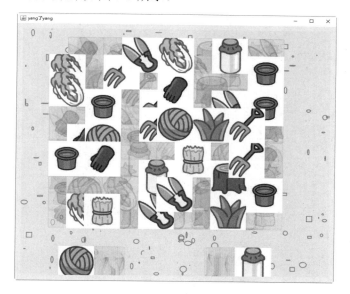

图 5-5　游戏界面

第 6 章　智能运动健身系统

　　生命在于运动，随着人们生活水平的提高，大家愈发重视自己的身体健康。为了提高自身的身体素质，越来越多的人开始参加健身运动。随着移动智能设备的兴起和普及，让传感器应用开发获得了良好的舞台，也便于开发者开发出和运动相关的智能设备。本章将通过一个综合实例的实现过程，详细讲解利用传感器技术开发运动轨迹计步器系统的方法，展示 Java 语言在移动智能设备中的应用。本章项目由 Android+传感器+Email+地图实现。

6.1 背景介绍

在现阶段，运动健身成为健康生活的新标配，跑步、撸铁成为许多人的锻炼方式。随着社会经济的快速发展，我国城镇化率逐步提高，城市人口日渐增多，人们也更加注重生活质量的提高，对个人健康以及身体塑形的关注度逐步增加，作为大健康产业之一的健身行业发展向好。2020年我国健身行业受到较大冲击，不过，伴随着政府积极出台相应支持措施，目前我国健身行业已经开始复苏。在"十四五"期间，我国为了建设体育强国，广泛开展全民健身运动，增强人民体质，在此背景下，健身行业迎来新一轮发展。

近日，运动科技平台 Keep 联合人民健康发布《2022 国民健身趋势报告》。报告由国家体育总局群众体育司指导，从运动人群画像、运动行为习惯等多个维度对国民的运动趋势进行了统计和分析。报告显示，近年来，国民正呈现出积极健身的趋势，表现为养成健身习惯人数增多，主动健身意识增强等。随着国民健身意识的觉醒，运动健身相关产品与服务将迎来高需求，同时"健身热"或将持续升温，成为促进体育消费产业化的重要助推力量。

6.2 运动健身发展趋势

国内线上线下健身的加速融合，推动着体育产业的转型与升级，形成了"互联网+健身"这种新的商业模式与业态，运动健身行业焕发出新的活力，行业整体将迎来二次发展。国家和企业积极推动健身机构社区化、线上化模式的实施，帮助健身机构进一步升级，为客户提供更为人性化、专业性的服务。

扫码看视频

在移动互联网的各类应用不断丰富之际，以健身 App 为特征的互联网健身也受到越来越多健身达人的青睐。与此同时，作为新的行业风口，相关企业也纷纷发力，加快可穿戴设备等硬件的开发和线下布局，以智能科技与运动健身相结合的创新方式，推动着运动健身行业快速发展。

随着人工智能和物联网技术的发展，运动健身行业与人工智能技术的联系愈发紧密，变化主要集中于场景、产品与用户体验。目前人工智能技术在智能运动健身行业中得到了不同程度应用。在智能健身硬件上，人工智能可通过动作识别等方式对运动姿态给予纠正，同时还可通过分析用户的运动表现为其提供健身课程的个性化建议。物联网可通过基础层面的芯片、传感器、计算平台部分与技术层面的计算机视觉、语音识别和机器学习等技术，为用户全面提升智能健身的体验。

6.3 系统分析

本智能运动健身系统的功能是，通过 Android 设备中的传感器记录当前的运动信息，包括位置、速率、海拔、记录频率和距离等，并且能将运动轨迹信息打包上传或分享。

扫码看视频

6.3.1 技术分析

在 Android 系统中，需要使用加速度传感器、线性加速度传感器和距离传感器来检测设备的运动数据。在当前的技术条件下，距离传感器是指利用飞行时间法(flying time)的原理来测量距离，以实现检测物体距离。飞行时间法是通过发射特别短的光脉冲，并测量此光脉冲从发射到被物体反射回来的时间，通过此时间间隔来计算物体之间的距离。

距离传感器在智能手机中的应用比较常见。一般触屏智能手机在默认设置下，都会有一个延时锁屏的功能，就是在一段时间内手机检测不到任何操作，就会进入锁屏状态。这样做是有一定好处的。手机作为移动终端的一种，追求低功耗是设计的目标之一。延时锁屏既可以避免不必要的能量消耗，又能保证不丢失重要信息。另外，在使用触屏手机接电话时，距离传感器会起作用：脸靠近屏幕时，屏幕灯会熄灭并自动锁屏，这样可以防止误操作；当脸离开屏幕时，屏幕灯会自动开启并且自动解锁。

除了广泛应用于手机设备之外，距离传感器还被用于野外环境探测、飞机高度检测、矿井深度检测、物料高度测量等领域。在野外应用领域中，主要用于检测山体情况和峡谷深度等。而在飞机高度检测领域主要是通过检测飞机在起飞和降落时距离地面的高度，并将结果实时显示在控制面板上。也可以使用距离传感器测量物料各点高度，用于计算物料的体积。在显示应用中，用于飞机高度和物料高度检测的距离传感器有 LDM301 系列，用于野外应用的距离传感器有 LDM4x 系列。

在当前的设备应用中，距离传感器被应用于智能皮带。在皮带扣里嵌入距离传感器，当把皮带调整至合适宽度、卡好皮带扣后，如果皮带在 10 秒内没有重新解开，传感器就会自动生成腰围数据。皮带与皮带扣连接处的其中一枚铆钉可作为数据传输装置。当智能手机在铆钉处保持两秒静止，手机里的自我健康管理 App 会自动激活，并获取本次腰围数据。

6.3.2 模块分析

本章的智能运动健身系统的构成模块如图 6-1 所示。

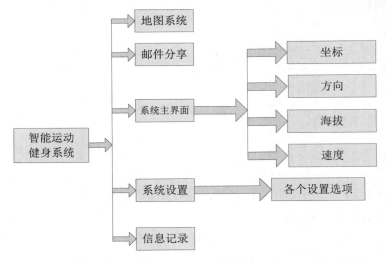

图 6-1 系统构成模块

6.4 系统主界面

系统主界面是运行程序后首先呈现在用户面前的界面。本节将详细讲解智能运动健身系统主界面的具体实现流程。

扫码看视频

6.4.1 布局文件

本系统主界面的布局文件是 main.xml，功能是通过文本控件显示当前的位置信息和传感器信息，主要实现代码如下。

```
<ScrollView xmlns:android="http://schemas.android.com/apk/res/android"
      android:id="@+id/scroll" android:layout_width="fill_parent"
      android:layout_height="fill_parent" android:background="#000000">
  <LinearLayout
      android:layout_width="fill_parent" android:layout_height="fill_parent"
      android:orientation="vertical">
    <TextView android:id="@+id/textStatus" android:layout_width="wrap_content"
      android:layout_height="wrap_content"/>
```

```xml
<TableLayout android:id="@+id/TableGPS"
        android:layout_width="fill_parent" android:layout_height="wrap_content"
        android:stretchColumns="1" android:background="#000000">
    <TableRow android:background="#333333" android:layout_margin="1dip">
        <TextView android:id="@+id/txtDateTimeAndProvider"
            android:gravity="left" android:textStyle="bold" android:padding="2dip"
            android:layout_span="2"/>
    </TableRow>
    <TableRow android:background="#333333" android:layout_margin="1dip">
        <TextView android:textStyle="bold" android:text="@string/txt_latitude"
            android:padding="3dip" android:textSize="17sp"/>
        <TextView android:id="@+id/txtLatitude" android:gravity="left"
            android:padding="3dip" android:textColor="#e8a317"
            android:textStyle="bold" android:textSize="18sp"/>
    </TableRow>
    <TableRow android:background="#333333" android:layout_margin="1dip">
        <TextView android:textStyle="bold" android:text="@string/txt_longitude"
            android:padding="3dip" android:textSize="17sp"/>
        <TextView android:id="@+id/txtLongitude" android:gravity="left"
            android:padding="3dip" android:textColor="#e8a317"
            android:textStyle="bold" android:textSize="18sp"/>
    </TableRow>
    <TableRow android:background="#333333" android:layout_margin="1dip">
        <TextView android:id="@+id/lblAltitude" android:textStyle="bold"
            android:text="@string/txt_altitude" android:padding="3dip"/>
        <TextView android:id="@+id/txtAltitude" android:gravity="left"
            android:padding="3dip"/>
    </TableRow>
    <TableRow android:background="#333333" android:layout_margin="1dip">
        <TextView android:id="@+id/lblSpeed" android:textStyle="bold"
            android:text="@string/txt_speed" android:padding="3dip"/>
        <TextView android:id="@+id/txtSpeed" android:gravity="left"
            android:padding="3dip"/>
    </TableRow>
    <TableRow android:background="#333333" android:layout_margin="1dip">
        <TextView android:id="@+id/lblDirection" android:textStyle="bold"
            android:text="@string/txt_direction" android:padding="3dip"/>
        <TextView android:id="@+id/txtDirection" android:gravity="left"
            android:padding="3dip"/>
    </TableRow>
    <TableRow android:background="#333333" android:layout_margin="1dip">
        <TextView android:id="@+id/lblSatellites" android:textStyle="bold"
            android:text="@string/txt_satellites" android:padding="3dip"/>
        <TextView android:id="@+id/txtSatellites" android:gravity="left"
            android:padding="3dip"/>
```

```xml
        </TableRow>
        <TableRow android:background="#333333" android:layout_margin="1dip">
            <TextView android:id="@+id/lblAccuracy" android:textStyle="bold"
                android:text="@string/txt_accuracy" android:padding="3dip"/>
            <TextView android:id="@+id/txtAccuracy" android:gravity="left"
                android:padding="3dip"/>
        </TableRow>
    </TableLayout>
…
```

6.4.2 实现主 Activity

在 Android 系统中，距离传感器也被称为 P-Sensor，值为 TYPE_PROXIMITY，单位为 cm，能够测量某个对象到屏幕的距离，例如，在打电话时判断人耳到电话屏幕的距离。

P-Sensor 主要用于在通话过程中防止用户误操作屏幕，接下来以通话过程为例来讲解电话程序对 P-Sensor 的操作流程。

(1) 在启动电话程序的时候，在 Java 文件中新建一个 P-Sensor 的 WakeLock 对象，例如以下所示的代码。

```
mProximityWakeLock = pm.newWakeLock(
PowerManager.PROXIMITY_SCREEN_OFF_WAKE_LOCK, LOG_TAG
);
```

对象 WakeLock 的功能是请求控制屏幕的点亮或熄灭。

(2) 在电话状态发生改变时，例如接通了电话，调用 Java 文件中的方法根据当前电话的状态来决定是否打开 P-Sensor。如果在通话过程中电话是 OFF-HOOK 状态，则打开 P-Sensor。例如下面的演示代码。

```
if (!mProximityWakeLock.isHeld()) {
    if (DBG) Log.d(LOG_TAG, "updateProximitySensorMode: acquiring...");
    mProximityWakeLock.acquire();
}
```

在上述代码中，mProximityWakeLock.acquire()会调用另外的方法打开 P-Sensor，这个方法会判断当前手机有没有 P-Sensor。如果有的话，就会向 SensorManager 注册一个 P-Sensor 监听器。这样当 P-Sensor 检测到手机和人体距离发生改变时，就会调用服务监听器进行处理。同样，当电话挂断时，电话模块会调用方法取消 P-Sensor 监听器。

本系统的主 Activity 是 GpsMainActivity，实现文件是 GpsMainActivity.java，具体实现流程如下。

(1) 定义更新 UI 线程的类 GpsMainActivity，获取 ToggleButton 按钮的开关来显示位置信息，具体实现代码如下。

```java
public class GpsMainActivity extends Activity implements OnCheckedChangeListener,
        IGpsLoggerServiceClient
{
    /**
     * 用处理器更新UI线程
     */
    public final Handler handler = new Handler();
    private static Intent serviceIntent;
    private GpsLoggingService loggingService;
    /**
     * 提供一个连接到GPS记录的服务
     */
    private ServiceConnection gpsServiceConnection = new ServiceConnection()
    {
        public void onServiceDisconnected(ComponentName name)
        {
            loggingService = null;
        }
        public void onServiceConnected(ComponentName name, IBinder service)
        {
            loggingService = ((GpsLoggingService.GpsLoggingBinder) service).getService();
            GpsLoggingService.SetServiceClient(GpsMainActivity.this);
            // 设置切换按钮，显示现有的位置信息
            ToggleButton buttonOnOff = (ToggleButton) findViewById(R.id.buttonOnOff);
            if (Session.isStarted())
            {
                buttonOnOff.setChecked(true);
                DisplayLocationInfo(Session.getCurrentLocationInfo());
            }
            buttonOnOff.setOnCheckedChangeListener(GpsMainActivity.this);
        }
    };
```

(2) 定义第一次创建样式时触发的方法 onCreate()，具体实现代码如下。

```java
    /**
     * 第一次创建样式时触发的事件
     */
    @Override
    public void onCreate(Bundle savedInstanceState){
        SharedPreferences prefs = PreferenceManager.getDefaultSharedPreferences(getBaseContext());
        String lang = prefs.getString("locale_override", "");
        if (!lang.equalsIgnoreCase(""))
        {
            Locale locale = new Locale(lang);
            Locale.setDefault(locale);
            Configuration config = new Configuration();
```

```
            config.locale = locale;
            getBaseContext().getResources().updateConfiguration(config,
                    getBaseContext().getResources().getDisplayMetrics());
        }
        super.onCreate(savedInstanceState);
        Utilities.LogInfo("GPSLogger started");
        setContentView(R.layout.main);
        GetPreferences();
        StartAndBindService();
    }
```

(3) 编写函数 StartAndBindService()，功能是启动定位服务并绑定到当前的 Activity 界面，具体实现代码如下。

```
private void StartAndBindService(){
    Utilities.LogDebug("StartAndBindService - binding now");
    serviceIntent = new Intent(this, GpsLoggingService.class);
    // Start the service in case it isn't already running
    startService(serviceIntent);
    // Now bind to service
    bindService(serviceIntent, gpsServiceConnection, Context.BIND_AUTO_CREATE);
    Session.setBoundToService(true);
}
```

(4) 编写函数 StopAndUnbindServiceIfRequired()，功能是单击关闭按钮时停止系统的监听服务，具体实现代码如下。

```
private void StopAndUnbindServiceIfRequired(){
    if(Session.isBoundToService())
    {
        unbindService(gpsServiceConnection);
        Session.setBoundToService(false);
    }
    if(!Session.isStarted())
    {
        Utilities.LogDebug("StopServiceIfRequired - Stopping the service");
        stopService(serviceIntent);
    }
}

@Override
protected void onPause(){
    StopAndUnbindServiceIfRequired();
    super.onPause();
}
@Override
protected void onDestroy(){
```

```
        StopAndUnbindServiceIfRequired();
        super.onDestroy();

    }
```

(5) 当单击切换按钮时调用函数 onCheckedChanged()，具体实现代码如下。

```
public void onCheckedChanged(CompoundButton buttonView, boolean isChecked){
    if (isChecked){
    GetPreferences();

        loggingService.StartLogging();
    }
    else{
        loggingService.StopLogging();
    }
}
```

(6) 编写函数 ShowPreferencesSummary()，功能是根据用户设置选项的值显示一个具有良好可读性的视图界面。主要实现代码如下。

```
private void ShowPreferencesSummary()
{
    TextView txtLoggingTo = (TextView) findViewById(R.id.txtLoggingTo);
    TextView txtFrequency = (TextView) findViewById(R.id.txtFrequency);
    TextView txtDistance = (TextView) findViewById(R.id.txtDistance);
    TextView txtAutoEmail = (TextView) findViewById(R.id.txtAutoEmail);
    if (!AppSettings.shouldLogToKml() && !AppSettings.shouldLogToGpx())
    {
        txtLoggingTo.setText(R.string.summary_loggingto_screen);
    }
    else if (AppSettings.shouldLogToGpx() && AppSettings.shouldLogToKml())
    {
        txtLoggingTo.setText(R.string.summary_loggingto_both);
    }
    else
    {
        txtLoggingTo.setText((AppSettings.shouldLogToGpx() ? "GPX" : "KML"));
    }
    if (AppSettings.getMinimumSeconds() > 0)
    {
        String descriptiveTime =
                Utilities.GetDescriptiveTimeString(AppSettings.getMinimumSeconds(),
                    getBaseContext());
        txtFrequency.setText(descriptiveTime);
    }
    else
    {
```

```java
            txtFrequency.setText(R.string.summary_freq_max);
    }
    if (AppSettings.getMinimumDistance() > 0)
    {
        if (AppSettings.shouldUseImperial())
        {
            int minimumDistanceInFeet =
                Utilities.MetersToFeet(AppSettings.getMinimumDistance());
            txtDistance.setText(((minimumDistanceInFeet == 1)
                ? getString(R.string.foot)
                : String.valueOf(minimumDistanceInFeet) + getString(R.string.feet)));
        }
        else
        {
            txtDistance.setText(((AppSettings.getMinimumDistance() == 1)
                ? getString(R.string.meter)
                : String.valueOf(AppSettings.getMinimumDistance()) +
                            getString(R.string.meters)));
        }
    }
    else
    {
        txtDistance.setText(R.string.summary_dist_regardless);
    }
    if (AppSettings.isAutoEmailEnabled())
    {
        String autoEmailResx;
        if (AppSettings.getAutoEmailDelay() == 0)
        {
            autoEmailResx = "autoemail_frequency_whenistop";
        }
        else
        {

            autoEmailResx = "autoemail_frequency_"
                + String.valueOf(AppSettings.getAutoEmailDelay()).replace(".", "");
            //.replace(".0", "")
        }
        String autoEmailDesc = getString(getResources().getIdentifier(autoEmailResx,
                        "string", getPackageName()));
        txtAutoEmail.setText(autoEmailDesc);
    }
    else
    {
        TableRow trAutoEmail = (TableRow) findViewById(R.id.trAutoEmail);
        trAutoEmail.setVisibility(View.INVISIBLE);
    }
}
```

(7) 编写函数 onOptionsItemSelected(),功能是监听用户选择的菜单项,然后根据所选的菜单项调用不同的处理方法。具体实现代码如下。

```java
public boolean onOptionsItemSelected(MenuItem item) {
    int itemId = item.getItemId();
    Utilities.LogInfo("Option item selected - " + String.valueOf(item.getTitle()));
    switch (itemId){
        case R.id.mnuSettings:
            Intent settingsActivity = new Intent(getBaseContext(),
                                GpsSettingsActivity.class);
            startActivity(settingsActivity);
            break;
        case R.id.mnuOSM:
            UploadToOpenStreetMap();
            break;
        case R.id.mnuAnnotate:
            Annotate();
            break;
        case R.id.mnuShare:
            Share();
            break;
        case R.id.mnuEmailnow:
            EmailNow();
            break;
        case R.id.mnuExit:
            loggingService.StopLogging();
            loggingService.stopSelf();
            System.exit(0);
            break;
    }
    return false;
}
private void EmailNow()
{
    if(Utilities.IsEmailSetup(getBaseContext())){
        loggingService.ForceEmailLogFile();
    }
    else{
        Intent emailSetup = new Intent(getBaseContext(), AutoEmailActivity.class);
        startActivity(emailSetup);
    }
}
```

(8) 通过方法 Share()实现轨迹分享功能,允许用户发送带位置的 GPX/KML 文件,在发送信息时可以选择一种分享方式,可以使用的分享方式有 Facebook(脸书)、短信、电子邮件、Twitte(推特)和蓝牙。具体实现代码如下。

```java
private void Share(){
    try{
        final String locationOnly = getString(R.string.sharing_location_only);
        final File gpxFolder = new File(Environment.getExternalStorageDirectory(),
                            "GPSLogger");
        if (gpxFolder.exists())
        {
            String[] enumeratedFiles = gpxFolder.list();
            List<String> fileList = new
                    ArrayList<String>(Arrays.asList(enumeratedFiles));
            Collections.reverse(fileList);
            fileList.add(0, locationOnly);
            final String[] files = fileList.toArray(new String[0]);
            final Dialog dialog = new Dialog(this);
            dialog.setTitle(R.string.sharing_pick_file);
            dialog.setContentView(R.layout.filelist);
            ListView thelist = (ListView) dialog.findViewById(R.id.listViewFiles);
            thelist.setAdapter(new ArrayAdapter<String>(getBaseContext(),
                    android.R.layout.simple_list_item_single_choice, files));
            thelist.setOnItemClickListener(new OnItemClickListener()
            {
                public void onItemClick(AdapterView<?> av, View v, int index, long arg)
                {
                    dialog.dismiss();
                    String chosenFileName = files[index];
                    final Intent intent = new Intent(Intent.ACTION_SEND);
                    // intent.setType("text/plain");
                    intent.setType("*/*");
                    if (chosenFileName.equalsIgnoreCase(locationOnly))
                    {
                        intent.setType("text/plain");
                    }
                    intent.putExtra(Intent.EXTRA_SUBJECT,
                            getString(R.string.sharing_mylocation));
                    if (Session.hasValidLocation())
                    {
                        String bodyText = getString(R.string.sharing_latlong_text,
                                String.valueOf(Session.getCurrentLatitude()),
                                String.valueOf(Session.getCurrentLongitude()));
                            intent.putExtra(Intent.EXTRA_TEXT, bodyText);
                            intent.putExtra("sms_body", bodyText);
                    }
                    if (chosenFileName.length() > 0
                            && !chosenFileName.equalsIgnoreCase(locationOnly))
                    {
                        intent.putExtra(Intent.EXTRA_STREAM,
                            Uri.fromFile(new File(gpxFolder, chosenFileName)));
```

```
                    }
                    startActivity(Intent.createChooser(intent,
                                getString(R.string.sharing_via)));
                }
            });
            dialog.show();
        }
        else
        {
            Utilities.MsgBox(getString(R.string.sorry),
                        getString(R.string.no_files_found), this);
        }
}
```

(9) 编写方法 UploadToOpenStreetMap()，功能是上传一个跟踪 GPS 记录的对象，具体实现代码如下。

```
private void UploadToOpenStreetMap()
{
    if(!Utilities.IsOsmAuthorized(getBaseContext()))
    {
        startActivity(Utilities.GetOsmSettingsIntent(getBaseContext()));
        return;
    }
    final String goToOsmSettings = getString(R.string.menu_settings);

    final File gpxFolder = new File(Environment.getExternalStorageDirectory(),
                        "GPSLogger");
    if (gpxFolder.exists())
    {
        FilenameFilter select = new FilenameFilter()
        {

            public boolean accept(File dir, String filename)
            {
            return filename.toLowerCase().contains(".gpx");
        }
        };
        String[] enumeratedFiles = gpxFolder.list(select);
        List<String> fileList = new
                    ArrayList<String>(Arrays.asList(enumeratedFiles));
        Collections.reverse(fileList);
        fileList.add(0, goToOsmSettings);
        final String[] files = fileList.toArray(new String[0]);
        final Dialog dialog = new Dialog(this);
        dialog.setTitle(R.string.osm_pick_file);
        dialog.setContentView(R.layout.filelist);
```

```
            ListView thelist = (ListView) dialog.findViewById(R.id.listViewFiles);
            thelist.setAdapter(new ArrayAdapter<String>(getBaseContext(),
                    android.R.layout.simple_list_item_single_choice, files));
            thelist.setOnItemClickListener(new OnItemClickListener()
            {
                public void onItemClick(AdapterView<?> av, View v, int index, long arg)
                {

                    dialog.dismiss();
                    String chosenFileName = files[index];

                    if(chosenFileName.equalsIgnoreCase(goToOsmSettings))
                    {
                        startActivity(Utilities.GetOsmSettingsIntent(getBaseContext()));
                    }
                    else
                    {
                        OSMHelper osm = new OSMHelper(GpsMainActivity.this);
                        Utilities.ShowProgress(GpsMainActivity.this,
                    getString(R.string.osm_uploading), getString(R.string.please_wait));
                        osm.UploadGpsTrace(chosenFileName);
                    }
                }
            });
            dialog.show();
        }
        else
        {
            Utilities.MsgBox(getString(R.string.sorry),
                    getString(R.string.no_files_found), this);
        }
    }
```

（10）编写函数 Annotate()，功能是提示用户输入信息，然后添加文本日志文件，具体实现代码如下。

```
    private void Annotate()
    {
        if (!AppSettings.shouldLogToGpx() && !AppSettings.shouldLogToKml())
        {
            return;
        }

        if (!Session.shoulAllowDescription())
        {
            Utilities.MsgBox(getString(R.string.not_yet),
                    getString(R.string.cant_add_description_until_next_point),
```

```
                            GetActivity());
            return;
    }
    AlertDialog.Builder alert = new AlertDialog.Builder(GpsMainActivity.this);
    alert.setTitle(R.string.add_description);
    alert.setMessage(R.string.letters_numbers);
    //新建一个EditText视图,用来获取用户的输入
    final EditText input = new EditText(getBaseContext());
    alert.setView(input);
    alert.setPositiveButton(R.string.ok, new DialogInterface.OnClickListener()
    {
        public void onClick(DialogInterface dialog, int whichButton)
        {
            final String desc = Utilities.CleanDescription(input.getText().toString());
            Annotate(desc);
        }
    });
    alert.setNegativeButton(R.string.cancel, new DialogInterface.OnClickListener()
    {
        public void onClick(DialogInterface dialog, int whichButton)
        {
            // Cancelled.
        }
    });
    alert.show();
}
```

(11) 编写函数 ClearForm(),功能是清理当前屏幕视图,并删除所有获取的值。具体实现代码如下。

```
public void ClearForm()
{
    TextView tvLatitude = (TextView) findViewById(R.id.txtLatitude);
    TextView tvLongitude = (TextView) findViewById(R.id.txtLongitude);
    TextView tvDateTime = (TextView) findViewById(R.id.txtDateTimeAndProvider);
    TextView tvAltitude = (TextView) findViewById(R.id.txtAltitude);
    TextView txtSpeed = (TextView) findViewById(R.id.txtSpeed);
    TextView txtSatellites = (TextView) findViewById(R.id.txtSatellites);
    TextView txtDirection = (TextView) findViewById(R.id.txtDirection);
    TextView txtAccuracy = (TextView) findViewById(R.id.txtAccuracy);
    tvLatitude.setText("");
    tvLongitude.setText("");
    tvDateTime.setText("");
    tvAltitude.setText("");
    txtSpeed.setText("");
    txtSatellites.setText("");
    txtDirection.setText("");
```

```
        txtAccuracy.setText("");
    }
```

(12) 编写函数 SetStatus()，功能是设置顶部状态标签的信息，具体实现代码如下。

```java
private void SetStatus(String message)
{
    TextView tvStatus = (TextView) findViewById(R.id.textStatus);
    tvStatus.setText(message);
    Utilities.LogInfo(message);
}
```

(13) 编写函数 SetSatelliteInfo()，功能是设置表中的卫星视图，具体实现代码如下。

```java
private void SetSatelliteInfo(int number)
{
    Session.setSatelliteCount(number);
    TextView txtSatellites = (TextView) findViewById(R.id.txtSatellites);
    txtSatellites.setText(String.valueOf(number));
}
```

(14) 编写函数 DisplayLocationInfo()，功能是处理指定的位置坐标，并将结果显示在位置视图中。具体实现代码如下。

```java
private void DisplayLocationInfo(Location loc)
{
    try
    {
        if (loc == null)
        {
            return;
        }
        Session.setLatestTimeStamp(System.currentTimeMillis());
        TextView tvLatitude = (TextView) findViewById(R.id.txtLatitude);
        TextView tvLongitude = (TextView) findViewById(R.id.txtLongitude);
        TextView tvDateTime = (TextView) findViewById(R.id.txtDateTimeAndProvider);
        TextView tvAltitude = (TextView) findViewById(R.id.txtAltitude);
        TextView txtSpeed = (TextView) findViewById(R.id.txtSpeed);
        TextView txtSatellites = (TextView) findViewById(R.id.txtSatellites);
        TextView txtDirection = (TextView) findViewById(R.id.txtDirection);
        TextView txtAccuracy = (TextView) findViewById(R.id.txtAccuracy);
        String providerName = loc.getProvider();
        if (providerName.equalsIgnoreCase("gps"))
        {
            providerName = getString(R.string.providername_gps);
        }
        else
        {
```

```
            providerName = getString(R.string.providername_celltower);
        }
        tvDateTime.setText(new Date().toLocaleString()
                + getString(R.string.providername_using, providerName));
        tvLatitude.setText(String.valueOf(loc.getLatitude()));
        tvLongitude.setText(String.valueOf(loc.getLongitude()));
        if (loc.hasAltitude())
        {
            double altitude = loc.getAltitude();
            if (AppSettings.shouldUseImperial())
            {
                tvAltitude.setText(String.valueOf(Utilities.MetersToFeet(altitude))
                        + getString(R.string.feet));
            }
            else
            {
                tvAltitude.setText(String.valueOf(altitude) + getString(R.string.meters));
            }
        }
        else
        {
            tvAltitude.setText(R.string.not_applicable);
        }       if (loc.hasSpeed())
        {
            float speed = loc.getSpeed();
            if (AppSettings.shouldUseImperial())
            {
                txtSpeed.setText(String.valueOf(Utilities.MetersToFeet(speed))
                        + getString(R.string.feet_per_second));
            }
            else
            {
                txtSpeed.setText(String.valueOf(speed) +
                            getString(R.string.meters_per_second));
            }
        }
        else
        {
            txtSpeed.setText(R.string.not_applicable);
        }
        if (loc.hasBearing())
        {
            float bearingDegrees = loc.getBearing();
            String direction;
            direction = Utilities.GetBearingDescription(bearingDegrees, getBaseContext());
            txtDirection.setText(direction + "(" + String.valueOf
                            (Math.round(bearingDegrees))
                    + getString(R.string.degree_symbol) + ")");
        }
```

```
            else
            {
                txtDirection.setText(R.string.not_applicable);
            }
            if (!Session.isUsingGps())
            {
                txtSatellites.setText(R.string.not_applicable);
                Session.setSatelliteCount(0);
            }
            if (loc.hasAccuracy())
            {
                float accuracy = loc.getAccuracy();
                if (AppSettings.shouldUseImperial())
                {
                    txtAccuracy.setText(getString(R.string.accuracy_within,
                            String.valueOf(Utilities.MetersToFeet(accuracy)),
                                    getString(R.string.feet)));
                }
                else
                {
                    txtAccuracy.setText(getString(R.string.accuracy_within,
                                String.valueOf(accuracy),
                            getString(R.string.meters)));
                }
            }
            else
            {
                txtAccuracy.setText(R.string.not_applicable);
            }
        }
        catch (Exception ex)
        {
            SetStatus(getString(R.string.error_displaying, ex.getMessage()));
        }
    }
```

6.4.3 系统服务

在主 Activity 的实现过程中用到了系统服务 Activity, 其实现文件是 GpsLoggingService.java, 功能是提供本系统所需要的后台服务方法。文件 GpsLoggingService.java 的具体实现流程如下。

(1) 定义可以调用的类和方法, 具体实现代码如下。

```
class GpsLoggingBinder extends Binder
{
    public GpsLoggingService getService()
```

```
            {
                Utilities.LogDebug("GpsLoggingBinder.getService");
                return GpsLoggingService.this;
            }
        }
```

(2) 编写函数 SetupAutoEmailTimers()，功能是建立基于用户偏好设置的电子邮件自动计时器，具体实现代码如下。

```
    private void SetupAutoEmailTimers()
    {
        Utilities.LogDebug("GpsLoggingService.SetupAutoEmailTimers");
        Utilities.LogDebug("isAutoEmailEnabled - " +
                        String.valueOf(AppSettings.isAutoEmailEnabled()));
        Utilities.LogDebug("Session.getAutoEmailDelay - " +
                        String.valueOf(Session.getAutoEmailDelay()));
        if (AppSettings.isAutoEmailEnabled() && Session.getAutoEmailDelay() > 0)
        {
            Utilities.LogDebug("Setting up email alarm");
            long triggerTime = System.currentTimeMillis()
                    + (long) (Session.getAutoEmailDelay() * 60 * 60 * 1000);
            alarmIntent = new Intent(getBaseContext(), AlarmReceiver.class);
            PendingIntent sender = PendingIntent.getBroadcast(this, 0, alarmIntent,
                    PendingIntent.FLAG_UPDATE_CURRENT);
            AlarmManager am = (AlarmManager) getSystemService(ALARM_SERVICE);
            am.set(AlarmManager.RTC_WAKEUP, triggerTime, sender);
        }
        else
        {
            Utilities.LogDebug("Checking if alarmIntent is null");
            if (alarmIntent != null)
            {
                Utilities.LogDebug("alarmIntent was null, canceling alarm");
                CancelAlarm();
            }
        }
    }
```

(3) 编写函数 AutoEmailLogFileOnStop()，如果用户选择使用自动邮件日志功能，则停止自动记录功能，具体实现代码如下。

```
    private void AutoEmailLogFileOnStop()
    {
      Utilities.LogDebug("GpsLoggingService.AutoEmailLogFileOnStop");
      Utilities.LogVerbose("isAutoEmailEnabled - " + AppSettings.isAutoEmailEnabled());
        // autoEmailDelay 0 means send it when you stop logging.
        if (AppSettings.isAutoEmailEnabled() && Session.getAutoEmailDelay() == 0)
```

```
        {
            Session.setEmailReadyToBeSent(true);
            AutoEmailLogFile();
        }
    }
```

(4) 编写函数 AutoEmailLogFile(),功能是调用自动电子邮件辅助处理文件并将其发送,具体实现代码如下。

```
private void AutoEmailLogFile()
{
    Utilities.LogDebug("GpsLoggingService.AutoEmailLogFile");
    Utilities.LogVerbose("isEmailReadyToBeSent - " + Session.isEmailReadyToBeSent());
    if (Session.getCurrentFileName() != null && Session.getCurrentFileName().length() > 0
            && Session.isEmailReadyToBeSent())
    {
        if(IsMainFormVisible())
        {
            Utilities.ShowProgress(mainServiceClient.GetActivity(),
                            getString(R.string.autoemail_sending),
                    getString(R.string.please_wait));
        }
        Utilities.LogInfo("Emailing Log File");
        AutoEmailHelper aeh = new AutoEmailHelper(GpsLoggingService.this);
        aeh.SendLogFile(Session.getCurrentFileName(), false);
        SetupAutoEmailTimers();

        if(IsMainFormVisible())
        {
            Utilities.HideProgress();
        }
    }
}
protected void ForceEmailLogFile()
{

    Utilities.LogDebug("GpsLoggingService.ForceEmailLogFile");
    if (Session.getCurrentFileName() != null && Session.getCurrentFileName().length() > 0)
    {
        if(IsMainFormVisible())
        {
            Utilities.ShowProgress(mainServiceClient.GetActivity(),
                            getString(R.string.autoemail_sending),
                    getString(R.string.please_wait));
        }
        Utilities.LogInfo("Force emailing Log File");
        AutoEmailHelper aeh = new AutoEmailHelper(GpsLoggingService.this);
```

```
        aeh.SendLogFile(Session.getCurrentFileName(), true);

        if(IsMainFormVisible())
        {
            Utilities.HideProgress();
        }
    }
}
```

(5) 编写函数 SetServiceClient()，功能是设置服务的活动形式，开启活动形式功能需要调用 IGpsLoggerServiceClient 服务。具体实现代码如下。

```
protected static void SetServiceClient(IGpsLoggerServiceClient mainForm)
{
    mainServiceClient = mainForm;
}
```

(6) 编写函数 GetPreferences()，功能是根据用户的偏好设置选择并填充 AppSettings 对象，再设置电子邮件的定时器。具体实现代码如下。

```
private void GetPreferences()
{
    Utilities.LogDebug("GpsLoggingService.GetPreferences");
    Utilities.PopulateAppSettings(getBaseContext());
    Utilities.LogDebug("Session.getAutoEmailDelay: " + Session.getAutoEmailDelay());
    Utilities.LogDebug("AppSettings.getAutoEmailDelay: " +
                    AppSettings.getAutoEmailDelay());
    if (Session.getAutoEmailDelay() != AppSettings.getAutoEmailDelay())
    {
        Utilities.LogDebug("Old autoEmailDelay - " +
                        String.valueOf(Session.getAutoEmailDelay())
                + "; New -" + String.valueOf(AppSettings.getAutoEmailDelay()));
        Session.setAutoEmailDelay(AppSettings.getAutoEmailDelay());
        SetupAutoEmailTimers();
    }
}
```

(7) 编写函数 StartLogging()，实现复位处理功能，具体实现代码如下。

```
protected void StartLogging()
{
    Utilities.LogDebug("GpsLoggingService.StartLogging");
    Session.setAddNewTrackSegment(true);
    if (Session.isStarted())
    {
        return;
    }
    Utilities.LogInfo("Starting logging procedures");
```

```
    startForeground(NOTIFICATION_ID, null);
    Session.setStarted(true);
    GetPreferences();
    Notify();
    ResetCurrentFileName();
    ClearForm();
    StartGpsManager();
}
```

(8) 编写函数 StopLogging(),功能是停止记录并删除通知,然后停止 GPS 服务,通过定时器停止邮件。具体实现代码如下。

```
protected void StopLogging()
{
    Utilities.LogDebug("GpsLoggingService.StopLogging");
    Session.setAddNewTrackSegment(true);
    Utilities.LogInfo("Stopping logging");
    Session.setStarted(false);
    AutoEmailLogFileOnStop();
    CancelAlarm();
    Session.setCurrentLocationInfo(null);
    stopForeground(true);
    RemoveNotification();
    StopGpsManager();
 StopMainActivity();
}
```

(9) 编写函数 Notify(),功能是在状态栏中显示通知,具体实现代码如下。

```
private void Notify()
{
    Utilities.LogDebug("GpsLoggingService.Notify");
    if (AppSettings.shouldShowInNotificationBar())
    {
        gpsNotifyManager = (NotificationManager) getSystemService(NOTIFICATION_SERVICE);
        ShowNotification();
    }
    else
    {
        RemoveNotification();
    }
}
```

(10) 编写函数 RemoveNotification(),若图标可见,则隐藏状态栏中的通知,具体实现代码如下。

```
private void RemoveNotification()
{
```

```
    Utilities.LogDebug("GpsLoggingService.RemoveNotification");
    try
    {
        if (Session.isNotificationVisible())
        {
            gpsNotifyManager.cancelAll();
        }
    }
    catch (Exception ex)
    {
        Utilities.LogError("RemoveNotification", ex);
    }
    finally
    {
        // notificationVisible = false;
        Session.setNotificationVisible(false);
    }
}
```

(11) 编写函数 ShowNotification()，功能是在状态栏中显示 GPS 记录器的通知图标，具体实现代码如下。

```
private void ShowNotification()
{
    Utilities.LogDebug("GpsLoggingService.ShowNotification");
    Intent contentIntent = new Intent(this, GpsMainActivity.class);
    PendingIntent pending = PendingIntent.getActivity(getBaseContext(), 0,
            contentIntent,android.content.Intent.FLAG_ACTIVITY_NEW_TASK);
    Notification nfc = new Notification(R.drawable.gpsloggericon2, null,
                    System.currentTimeMillis());
    nfc.flags |= Notification.FLAG_ONGOING_EVENT;
    NumberFormat nf = new DecimalFormat("###.######");
    String contentText = getString(R.string.gpslogger_still_running);
    if (Session.hasValidLocation())
    {
        contentText = nf.format(Session.getCurrentLatitude()) + ","
                + nf.format(Session.getCurrentLongitude());
    }
    nfc.setLatestEventInfo(getBaseContext(), getString(R.string.gpslogger_still_running),
            contentText, pending);
    gpsNotifyManager.notify(NOTIFICATION_ID, nfc);
    Session.setNotificationVisible(true);
}
```

(12) 编写函数 StartGpsManager()，根据用户的偏好设置选项启动 GPS 功能，具体实现代码如下。

```
private void StartGpsManager()
{
    Utilities.LogDebug("GpsLoggingService.StartGpsManager");
    GetPreferences();
    gpsLocationListener = new GeneralLocationListener(this);
    towerLocationListener = new GeneralLocationListener(this);
    gpsLocationManager = (LocationManager) getSystemService(Context.LOCATION_SERVICE);
    towerLocationManager = (LocationManager) getSystemService(Context.LOCATION_SERVICE);
    CheckTowerAndGpsStatus();
    if (Session.isGpsEnabled() && !AppSettings.shouldPreferCellTower())
    {
        Utilities.LogInfo("Requesting GPS location updates");
        gpsLocationManager.requestLocationUpdates(LocationManager.GPS_PROVIDER,
                AppSettings.getMinimumSeconds() * 1000,
                        AppSettings.getMinimumDistance(),
                gpsLocationListener);
        gpsLocationManager.addGpsStatusListener(gpsLocationListener);
        Session.setUsingGps(true);
    }
    else if (Session.isTowerEnabled())
    {
        Utilities.LogInfo("Requesting tower location updates");
        Session.setUsingGps(false);
        
towerLocationManager.requestLocationUpdates(LocationManager.NETWORK_PROVIDER,
                AppSettings.getMinimumSeconds() * 1000,
                        AppSettings.getMinimumDistance(),
                towerLocationListener);
    }
    else
    {
        Utilities.LogInfo("No provider available");
        Session.setUsingGps(false);
        SetStatus(R.string.gpsprovider_unavailable);
    SetFatalMessage(R.string.gpsprovider_unavailable);
    StopLogging();
        return;
    }
    SetStatus(R.string.started);
}
```

(13) 编写函数 CheckTowerAndGpsStatus(), 功能是周期性检查是否已经启动 GPS 和信号塔, 具体实现代码如下。

```
private void CheckTowerAndGpsStatus() {
Session.setTowerEnabled(towerLocationManager.
                isProviderEnabled(LocationManager.NETWORK_PROVIDER));
```

```
            Session.setGpsEnabled(gpsLocationManager.
                        isProviderEnabled(LocationManager.GPS_PROVIDER));
    }
```

(14) 编写函数 StopGpsManager(),功能是停止位置管理服务,具体实现代码如下。

```
    private void StopGpsManager()
    {
        Utilities.LogDebug("GpsLoggingService.StopGpsManager");
        if (towerLocationListener != null)
        {
            towerLocationManager.removeUpdates(towerLocationListener);
        }
        if (gpsLocationListener != null)
        {
            gpsLocationManager.removeUpdates(gpsLocationListener);
            gpsLocationManager.removeGpsStatusListener(gpsLocationListener);
        }
        SetStatus(getString(R.string.stopped));
    }
```

(15) 编写函数 ResetCurrentFileName(),功能是基于用户偏好设置当前文件名,具体实现代码如下。

```
    private void ResetCurrentFileName()
    {
        Utilities.LogDebug("GpsLoggingService.ResetCurrentFileName");
        String newFileName;
        if (AppSettings.shouldCreateNewFileOnceADay())
        {
            // 20100114.gpx
            SimpleDateFormat sdf = new SimpleDateFormat("yyyyMMdd");
            newFileName = sdf.format(new Date());
            Session.setCurrentFileName(newFileName);
        }
        else
        {
            // 20100114183326.gpx
            SimpleDateFormat sdf = new SimpleDateFormat("yyyyMMddHHmmss");
            newFileName = sdf.format(new Date());
            Session.setCurrentFileName(newFileName);
        }
        if (IsMainFormVisible())
        {
            mainServiceClient.onFileName(newFileName);
        }
    }
```

6.5 系统设置

当单击图 6-3 中的 Settings 选项后，会弹出系统设置界面。在系统设置界面中，可以设置系统的常用选项参数。本节将详细讲解系统设置模块的实现过程。

扫码看视频

6.5.1 选项设置

编写文件 AppSettings.java，功能是根据用户选择的选项值来设置系统，例如设置保存为 GPX(GPS eXchange Format 的缩写，译为 GPS 交换格式，是一个 XML 格式，可以用来描述路点、轨迹、路程)格式数据文件或 KML(一种文件格式，用于在地球浏览器中显示地理数据，例如 Google 地球、Google 地图和谷歌手机地图)格式数据文件。文件 AppSettings.java 的具体实现代码如下。

```java
public class AppSettings extends Application
{
    //用户设置
    private static boolean useImperial = false;
    public static boolean shouldUseImperial()
    {
        return useImperial;
    }
    static void setUseImperial(boolean useImperial)
    {
        AppSettings.useImperial = useImperial;
    }
    /**
     * @return the 一天更新一个新文件
     */
    public static boolean shouldCreateNewFileOnceADay()
    {
        return newFileOnceADay;
    }
    static void setNewFileOnceADay(boolean newFileOnceADay)
    {
        AppSettings.newFileOnceADay = newFileOnceADay;
    }
    public static boolean shouldPreferCellTower()
    {
        return preferCellTower;
    }
//省略后面的代码
```

6.5.2 生成 GPX 和 KML 格式的文件

在系统设置界面中，可以指定一个文件来保存行走轨迹。本系统提供了两种保存轨迹的文件格式，分别是 GPX 和 KML。本节编写文件 Gpx10FileLogger.java，生成 GPX 格式的文件，具体实现代码如下。

```java
class Gpx10FileLogger implements IFileLogger
{
    private final static Object lock = new Object();
    private File gpxFile = null;
    private boolean useSatelliteTime = false;
    private boolean addNewTrackSegment;
    private int satelliteCount;

    Gpx10FileLogger(File gpxFile, boolean useSatelliteTime, boolean addNewTrackSegment,
                int satelliteCount)
    {
        this.gpxFile = gpxFile;
        this.useSatelliteTime = useSatelliteTime;
        this.addNewTrackSegment = addNewTrackSegment;
        this.satelliteCount = satelliteCount;
    }
    public void Write(Location loc) throws Exception
    {
        try
        {
            Date now;
            if (useSatelliteTime)
            {
                now = new Date(loc.getTime());
            }
            else
            {
                now = new Date();
            }
            String dateTimeString = Utilities.GetIsoDateTime(now);

            if (!gpxFile.exists())
            {
                gpxFile.createNewFile();
                FileOutputStream initialWriter = new FileOutputStream(gpxFile, true);
                BufferedOutputStream initialOutput = new
                                            BufferedOutputStream(initialWriter);
                String initialXml = "<?xml version=\"1.0\"?>"
```

```java
                    + "<gpx version=\"1.0\" creator=\"GPSLogger - 
                    http://gpslogger.mendhak.com/\" "
                    + "xmlns:xsi=\"http://www.w3.org/2001/XMLSchema-instance\" "
                    + "xmlns=\"http://www.topografix.com/GPX/1/0\" "
                    + "xsi:schemaLocation=\"http://www.topografix.com/GPX/1/0 "
                    + "http://www.topografix.com/GPX/1/0/gpx.xsd\">"
                    + "<time>" + dateTimeString + "</time>" + "<bounds />" + "<trk></trk></gpx>";
            initialOutput.write(initialXml.getBytes());
            initialOutput.flush();
            initialOutput.close();
    }
    DocumentBuilderFactory factory = DocumentBuilderFactory.newInstance();
    DocumentBuilder builder = factory.newDocumentBuilder();
    Document doc = builder.parse(gpxFile);
    Node trkSegNode;
    NodeList trkSegNodeList = doc.getElementsByTagName("trkseg");
    if(addNewTrackSegment || trkSegNodeList.getLength()==0)
    {
        NodeList trkNodeList = doc.getElementsByTagName("trk");
        trkSegNode = doc.createElement("trkseg");
        trkNodeList.item(0).appendChild(trkSegNode);
    }
    else
    {
        trkSegNode = trkSegNodeList.item(trkSegNodeList.getLength()-1);
    }
    Element trkptNode = doc.createElement("trkpt");
    Attr latAttribute = doc.createAttribute("lat");
    latAttribute.setValue(String.valueOf(loc.getLatitude()));
    trkptNode.setAttributeNode(latAttribute);
    Attr lonAttribute = doc.createAttribute("lon");
    lonAttribute.setValue(String.valueOf(loc.getLongitude()));
    trkptNode.setAttributeNode(lonAttribute);
    if(loc.hasAltitude())
    {
        Node eleNode = doc.createElement("ele");
        eleNode.appendChild(doc.createTextNode(String.valueOf(loc.getAltitude())));
        trkptNode.appendChild(eleNode);
    }
    Node timeNode = doc.createElement("time");
    timeNode.appendChild(doc.createTextNode(dateTimeString));
    trkptNode.appendChild(timeNode);
    trkSegNode.appendChild(trkptNode);
    if(loc.hasBearing())
    {
        Node courseNode = doc.createElement("course");
        courseNode.appendChild(doc.createTextNode(String.valueOf(loc.getBearing())));
```

```java
            trkptNode.appendChild(courseNode);
        }
        if(loc.hasSpeed())
        {
            Node speedNode = doc.createElement("speed");
            speedNode.appendChild(doc.createTextNode(String.valueOf(loc.getSpeed())));
            trkptNode.appendChild(speedNode);
        }
        Node srcNode = doc.createElement("src");
        srcNode.appendChild(doc.createTextNode(loc.getProvider()));
        trkptNode.appendChild(srcNode);
        if(Session.getSatelliteCount() > 0)
        {
            Node satNode = doc.createElement("sat");
            satNode.appendChild(doc.createTextNode(String.valueOf(satelliteCount)));
            trkptNode.appendChild(satNode);
        }
        String newFileContents = Utilities.GetStringFromNode(doc);
        synchronized(lock)
        {
            FileOutputStream fos = new FileOutputStream(gpxFile, false);
            fos.write(newFileContents.getBytes());
            fos.close();
        }
    }
    catch (Exception e)
    {
        Utilities.LogError("Gpx10FileLogger.Write", e);
        throw new Exception("Could not write to GPX file");
    }
}
public void Annotate(String description) throws Exception
{
    if (!gpxFile.exists())
    {
        return;
    }
    try
    {
        DocumentBuilderFactory factory = DocumentBuilderFactory.newInstance();
        DocumentBuilder builder = factory.newDocumentBuilder();
        Document doc = builder.parse(gpxFile);
        NodeList trkptNodeList = doc.getElementsByTagName("trkpt");
        Node lastTrkPt = trkptNodeList.item(trkptNodeList.getLength()-1);
        Node nameNode = doc.createElement("name");
        nameNode.appendChild(doc.createTextNode(description));
        lastTrkPt.appendChild(nameNode);
```

```
                Node descNode = doc.createElement("desc");
                descNode.appendChild(doc.createTextNode(description));
                lastTrkPt.appendChild(descNode);
                String newFileContents = Utilities.GetStringFromNode(doc);
                synchronized(lock)
                {
                    FileOutputStream fos = new FileOutputStream(gpxFile, false);
                    fos.write(newFileContents.getBytes());
                    fos.close();
                }
            }
            catch(Exception e)
            {
                Utilities.LogError("Gpx10FileLogger.Annotate", e);
                throw new Exception("Could not annotate GPX file");
            }
        }
    }
```

编写文件 Kml10FileLogger.java,生成 KML 格式文件,具体方法和编写文件 Gpx10FileLogger.java 类似,这里不再赘述。

6.6 邮件分享提醒

在系统设置模块中,可以通过设置系统邮件来分享行走轨迹信息。

6.6.1 基本邮箱设置

扫码看视频

编写文件 AutoEmailActivity.java,设置发送邮件的邮箱地址、密码、邮件服务器等信息,以实现邮件自动发送功能。文件 AutoEmailActivity.java 的主要实现代码如下。

```
public class AutoEmailActivity extends PreferenceActivity implements
        OnPreferenceChangeListener, IMessageBoxCallback, IAutoSendHelper,
        OnPreferenceClickListener
{
    private final Handler handler = new Handler();
    @Override
    public void onCreate(Bundle savedInstanceState)
    {
        super.onCreate(savedInstanceState);
        addPreferencesFromResource(R.xml.autoemailsettings);
        CheckBoxPreference chkEnabled = (CheckBoxPreference)
                                    findPreference("autoemail_enabled");
        chkEnabled.setOnPreferenceChangeListener(this);
```

```java
        ListPreference lstPresets = (ListPreference) findPreference("autoemail_preset");
        lstPresets.setOnPreferenceChangeListener(this);
        EditTextPreference txtSmtpServer = (EditTextPreference)
                                    findPreference("smtp_server");
        EditTextPreference txtSmtpPort = (EditTextPreference)
                                    findPreference("smtp_port");
        txtSmtpServer.setOnPreferenceChangeListener(this);
        txtSmtpPort.setOnPreferenceChangeListener(this);
        Preference testEmailPref = (Preference) findPreference("smtp_testemail");
        testEmailPref.setOnPreferenceClickListener(this);
    }
    public boolean onPreferenceClick(Preference preference)
    {
        if (!IsFormValid())
        {
            Utilities.MsgBox(getString(R.string.autoemail_invalid_form),
                    getString(R.string.autoemail_invalid_form_message),
                    AutoEmailActivity.this);
            return false;
        }
        Utilities.ShowProgress(this, getString(R.string.autoemail_sendingtest),
                    getString(R.string.please_wait));
        CheckBoxPreference chkUseSsl = (CheckBoxPreference) findPreference("smtp_ssl");
        EditTextPreference txtSmtpServer = (EditTextPreference)
                                    findPreference("smtp_server");
        EditTextPreference txtSmtpPort = (EditTextPreference)
                                    findPreference("smtp_port");
        EditTextPreference txtUsername = (EditTextPreference)
                                    findPreference("smtp_username");
        EditTextPreference txtPassword = (EditTextPreference)
                                    findPreference("smtp_password");
        EditTextPreference txtTarget = (EditTextPreference)
                                    findPreference("autoemail_target");
        AutoEmailHelper aeh = new AutoEmailHelper(null);
        aeh.SendTestEmail(txtSmtpServer.getText(), txtSmtpPort.getText(),
                txtUsername.getText(), txtPassword.getText(),
                chkUseSsl.isChecked(), txtTarget.getText(),
                AutoEmailActivity.this, AutoEmailActivity.this);
                return true;
    }
    private boolean IsFormValid()
    {
        CheckBoxPreference chkEnabled = (CheckBoxPreference)
                                    findPreference("autoemail_enabled");
        EditTextPreference txtSmtpServer = (EditTextPreference)
                                    findPreference("smtp_server");
```

```java
            EditTextPreference txtSmtpPort = (EditTextPreference)
                                        findPreference("smtp_port");
            EditTextPreference txtUsername = (EditTextPreference)
                                        findPreference("smtp_username");
            EditTextPreference txtPassword = (EditTextPreference)
                                        findPreference("smtp_password");
            EditTextPreference txtTarget = (EditTextPreference) findPreference("autoemail_target");
            if (chkEnabled.isChecked())
            {
                if (txtSmtpServer.getText() != null
                        && txtSmtpServer.getText().length() > 0
                        && txtSmtpPort.getText() != null
                        && txtSmtpPort.getText().length() > 0
                        && txtUsername.getText() != null
                        && txtUsername.getText().length() > 0
                        && txtPassword.getText() != null
                        && txtPassword.getText().length() > 0
                        && txtTarget.getText() != null
                        && txtTarget.getText().length() > 0)
                {
                    return true;
                }
                else
                {
                    return false;
                }
            }
            return true;
        }
        public boolean onKeyDown(int keyCode, KeyEvent event)
        {
            if (keyCode == KeyEvent.KEYCODE_BACK)
            {
                if (!IsFormValid())
                {
                    Utilities.MsgBox(getString(R.string.autoemail_invalid_form),
                            getString(R.string.autoemail_invalid_form_message),
                            this);
                    return false;
                }
                else
                {
                    return super.onKeyDown(keyCode, event);
                }
            }
            else
            {
```

```
                return super.onKeyDown(keyCode, event);
        }
    }
```

6.6.2 发送邮件

编写文件 AutoEmailHelper.java，功能是使用邮件设置模块的邮箱来发送邮件信息，主要实现代码如下。

```
public class AutoEmailHelper implements IAutoSendHelper
{
    private GpsLoggingService mainActivity;
    private boolean forcedSend = false;
    public AutoEmailHelper(GpsLoggingService activity)
    {
        this.mainActivity = activity;
    }
    public void SendLogFile(String currentFileName, boolean forcedSend)
    {
        this.forcedSend = forcedSend;
        Thread t = new Thread(new AutoSendHandler(currentFileName, this));
        t.start();
    }
    void SendTestEmail(String smtpServer, String smtpPort,
            String smtpUsername, String smtpPassword, boolean smtpUseSsl,
            String emailTarget, Activity callingActivity, IAutoSendHelper helper)
    {
        Thread t = new Thread(new TestEmailHandler(helper, smtpServer,
                smtpPort, smtpUsername, smtpPassword, smtpUseSsl, emailTarget));
        t.start();
    }
    public void OnRelay(boolean connectionSuccess, String errorMessage)
    {
        if (!connectionSuccess)
        {
            mainActivity.handler.post(mainActivity.updateResultsEmailSendError);
        }
        else
        {
            Utilities.LogInfo("Email sent");
            if (!forcedSend)
            {
                Utilities.LogDebug("setEmailReadyToBeSent = false");
                Session.setEmailReadyToBeSent(false);
            }
        }
    }
}
```

6.7 上传 OSM 地图

OSM 是 OpenStreetMap 的简称，这是一个网上地图协作计划，目标是创造一个内容自由且能让所有人编辑的世界地图。OSM 的地图数据是由用户贡献的，用户可以使用手持 GPS 设备、航空摄影照片、其他自由内容，甚至只依赖当地的智慧进行地图的绘制。网站这一开放的地图数据模式使得 OSM 成为一个众包地图项目，吸引了全球的贡献者，从而不断丰富和更新地图内容。OSM 网站的灵感来自维基百科等网站，注册的用户可上传 GPS 路径及使用内置的程序编辑数据。

扫码看视频

6.7.1 授权提示布局文件

授权提示界面的布局文件是 osmauth.xml，主要实现代码如下。

```xml
<TableLayout android:id="@+id/TableOSM"
    android:layout_width="fill_parent" android:layout_height="wrap_content"
    android:stretchColumns="1" android:background="#000000">
    <TableRow></TableRow>
    <TableRow>
    <TextView android:id="@+id/lblAuthorizeDescription"
android:layout_height="wrap_content"
        android:text="@string/osm_lbl_authorize_description"
android:layout_width="wrap_content"></TextView>
    </TableRow>
    <TableRow>
        <Button android:id="@+id/btnAuthorizeOSM"
            android:text="@string/osm_lbl_authorize"
android:layout_height="wrap_content" android:layout_width="wrap_content"/>
    </TableRow>
</TableLayout>
```

通过上述代码，可在屏幕中显示授权提示信息，并在屏幕下方显示一个激活按钮。单击激活按钮，会触发文件 OSMAuthorizationActivity.java，具体实现代码如下。

```java
public class OSMAuthorizationActivity extends Activity implements
        OnClickListener
{
    private static OAuthProvider provider;
    private static OAuthConsumer consumer;
    @Override
    public void onCreate(Bundle savedInstanceState)
    {
```

```java
        super.onCreate(savedInstanceState);
        setContentView(R.layout.osmauth);
        final Intent intent = getIntent();
        final Uri myURI = intent.getData();
        if (myURI != null && myURI.getQuery() != null
                && myURI.getQuery().length() > 0)
        {
            //用户已返回，从querystring中读取验证器信息
            String oAuthVerifier = myURI.getQueryParameter("oauth_verifier");
            try
            {
                SharedPreferences prefs =
                    PreferenceManager.getDefaultSharedPreferences(getBaseContext());

                if (provider == null)
                {
                    provider = Utilities.GetOSMAuthProvider(getBaseContext());
                }
                if (consumer == null)
                {
                    //如果使用者为null，则从存储的值重新初始化
                    consumer = Utilities.GetOSMAuthConsumer(getBaseContext());
                }
                //向OpenStreetMap询问访问令牌，这是主要事件
                provider.retrieveAccessToken(consumer, oAuthVerifier);

                String osmAccessToken = consumer.getToken();
                String osmAccessTokenSecret = consumer.getTokenSecret();
                SharedPreferences.Editor editor = prefs.edit();
                editor.putString("osm_accesstoken", osmAccessToken);
                editor.putString("osm_accesstokensecret", osmAccessTokenSecret);
                editor.commit();

                //Now go away
                startActivity(new Intent(getBaseContext(), GpsMainActivity.class));
                finish();
            }
            catch (Exception e)
            {
                Utilities.LogError("OSMAuthorizationActivity.onCreate - user has returned", e);
                Utilities.MsgBox(getString(R.string.sorry),
                        getString(R.string.osm_auth_error), this);
            }
        }
        Button authButton = (Button) findViewById(R.id.btnAuthorizeOSM);
        authButton.setOnClickListener(this);
    }
    public void onClick(View v)
    {
```

```
            try
            {
                    //用户点击,设置使用者consumer和提供者provider
                    consumer = Utilities.GetOSMAuthConsumer(getBaseContext());
                    provider = Utilities.GetOSMAuthProvider(getBaseContext());
                    String authUrl;
                    //获取请求令牌和请求令牌密钥
                    authUrl = provider.retrieveRequestToken(consumer, OAuth.OUT_OF_BAND);
                    //保存以备后续用
                    SharedPreferences prefs =
                            PreferenceManager.getDefaultSharedPreferences(getBaseContext());
                    SharedPreferences.Editor editor = prefs.edit();
                    editor.putString("osm_requesttoken", consumer.getToken());
                    editor.putString("osm_requesttokensecret",consumer.getTokenSecret());
                    editor.commit();
                    //打开浏览器,将用户信息发送到OpenStreetMap.org
                    Uri uri = Uri.parse(authUrl);
                    Intent intent = new Intent(Intent.ACTION_VIEW, uri);
                    startActivity(intent);
            }
            catch (Exception e)
            {
                    Utilities.LogError("OSMAuthorizationActivity.onClick", e);
                    Utilities.MsgBox(getString(R.string.sorry),
                                    getString(R.string.osm_auth_error), this);
            }
    }
}
```

6.7.2 文件上传

编写文件 OSMHelper.java,功能是在获取权限后上传 OpenStreetMap 轨迹文件,主要实现代码如下。

```
public class OSMHelper implements IOsmHelper
{
    private GpsMainActivity mainActivity;
    public OSMHelper(GpsMainActivity activity)
    {
        this.mainActivity = activity;
    }
    
    public void UploadGpsTrace(String fileName)
    {
        File gpxFolder = new File(Environment.getExternalStorageDirectory(), "GPSLogger");
        File chosenFile = new File(gpxFolder, fileName);
        OAuthConsumer consumer = Utilities.GetOSMAuthConsumer(mainActivity.getBaseContext());
        String gpsTraceUrl = mainActivity.getString(R.string.osm_gpstrace_url);
```

```java
        SharedPreferences prefs = PreferenceManager.getDefaultSharedPreferences
                        (mainActivity.getBaseContext());
        String description = prefs.getString("osm_description", "");
        String tags = prefs.getString("osm_tags", "");
        String visibility = prefs.getString("osm_visibility", "private");

        Thread t = new Thread(new OsmUploadHandler(this, consumer, gpsTraceUrl, chosenFile,
                description, tags, visibility));
        t.start();
    }
    public void OnComplete()
    {
        mainActivity.handler.post(mainActivity.updateOsmUpload);
    }
    private class OsmUploadHandler implements Runnable
    {
        OAuthConsumer consumer;
        String gpsTraceUrl;
        File chosenFile;
        String description;
        String tags;
        String visibility;
        IOsmHelper helper;

        public OsmUploadHandler(IOsmHelper helper, OAuthConsumer consumer, String
gpsTraceUrl, File chosenFile, String description, String tags, String visibility)
        {
            this.consumer = consumer;
            this.gpsTraceUrl = gpsTraceUrl;
            this.chosenFile = chosenFile;
            this.description = description;
            this.tags = tags;
            this.visibility = visibility;
            this.helper = helper;
        }

        public void run()
        {
            try
            {
              HttpPost request = new HttpPost(gpsTraceUrl);

            consumer.sign(request);
            MultipartEntity entity =
                        new MultipartEntity(HttpMultipartMode.BROWSER_COMPATIBLE);

            FileBody gpxBody = new FileBody(chosenFile);
            entity.addPart("file", gpxBody);
            if(description == null || description.length() <= 0)
```

```
{
        description = "GPSLogger for Android";
}

entity.addPart("description", new StringBody(description));
entity.addPart("tags", new StringBody(tags));
entity.addPart("visibility", new StringBody(visibility));

request.setEntity(entity);
DefaultHttpClient httpClient = new DefaultHttpClient();
HttpResponse response = httpClient.execute(request);
int statusCode = response.getStatusLine().getStatusCode();
Utilities.LogDebug("OSM Upload - " + String.valueOf(statusCode));
helper.OnComplete();

}
catch(Exception e)
{
        Utilities.LogError("OsmUploadHelper.run", e);
}
}
}
}
```

6.8 调试运行

至此，本项目主要功能的实现介绍完毕，系统主界面的执行效果如图 6-2 所示。单击设备中的 MENU 按钮后，在屏幕下方会弹出选项设置界面，如图 6-3 所示。

图 6-2 系统主界面

图 6-3 设置界面

第7章 图书市场数据分析系统

本章将通过一个综合实例的实现过程，讲解使用 Java 语言爬取某出版社图书数据的方法，并将爬取的图书数据保存到 MySQL 数据库中。然后使用 JSP 开发一个 Java Web 程序，在 Web 前端展示数据库中保存的爬虫数据，通过大数据技术分析图书的结构。本章项目由网络爬虫+JSP+MySQL+大数据分析+Echarts 实现。

7.1 图书市场介绍

精神文明是一个国家的灵魂，随着经济社会的发展，精神文明建设一步一步地发展，人们的文化素养和综合素质也随精神文明的发展而慢慢提高，同时，人们对精神文化的需求也越来越高。网络的发展让信息的传播越来越快，电子书的出现使我国的图书市场受到了一定的冲击，二者都有各自的优点：电子图书方便快捷，纸质图书则更具文化内涵。

扫码看视频

7.1.1 图书市场现状分析

《2022 年图书零售市场年度报告》显示，2022 年我国图书零售市场码洋规模为 871 亿元，短视频电商零售图书码洋同比上升 42.86%，码洋占比赶超实体书店，成为新书首发重要渠道。

该报告由北京开卷信息技术有限公司基于图书零售市场观测系统数据分析完成。据统计，2022 年图书零售市场较 2021 年同比下降 11.77%。从不同渠道零售图书市场看，实体店渠道零售图书码洋同比下降 37.22%，平台电商同比下降 16.06%，短视频电商实现正增长。从 2022 年各类图书的码洋构成看，少儿类图书码洋比重最大，且码洋比重进一步上升；教辅类图书码洋比重位居第二，文学类图书码洋比重增幅最大。

报告分析指出，出版行业作为以内容为核心的创意产业，要实现出版繁荣，不仅需要大量品种，还需要增强原创能力。2022 年，我国原创新书品种规模占比升高，文学类新书中原创图书占比最高。

7.1.2 图书市场背景分析

从经济方面来看，随着我国国民收入的增加，教育文化娱乐消费支出的增长，居民人均教育文化娱乐消费支出也随之增长。在 2022 年，我国人均教育文化娱乐消费支出占人均消费支出的比重为 10.8%，全国居民人均教育文化娱乐支出两年平均增长 1.7%。居民收入增长及文化消费意愿的增强，成为图书行业发展的重要推动力。

近年来，中央和地方先后出台一系列政策，对实体书店提供税收减免、项目补贴等扶持，在保障行业健康发展、推动转型升级、承担社会责任等方面取得了较好的效果。《出版业"十四五"时期发展规划》(下面简称《规划》)提出的目标是：服务大局的能力达到新高度，满足人民学习阅读需求实现新提升，行业繁荣发展取得新突破，产业数字化水平迈上新台阶，出版业走出去取得新成效和行业治理效能得到新提高。《规划》还提出要整理出版一批重要文化典籍，推出一批人文社科领域出版精品，推出一批科学技术类出版精品和一批少儿读物精品。

7.1.3 图书市场发展趋势

长期来看，与欧美发达国家相比，我国图书零售折扣较高，因此我国图书零售实际收入相较于码洋规模存在较大折扣，对比部分发达国家图书零售行业的实际收入规模，我国图书零售规模存在较大增长空间。另一方面，目前我国图书定价机制与发达国家存在差异，零售过程中折扣现象相对无序。未来随着行业监管的不断完善以及文化支出的提升，我国图书零售市场规模存在长期增长的可能。

据权威机构数据统计，我国图书出版行业的发展趋势如下。

(1) 销售渠道变革，线上销售占据主导地位。

渠道变革为图书出版行业带来诸多机遇与挑战。图书作为一种相对标准化的商品，在线上销售能够更广泛触达用户，也能够一定程度降低实体书店的运营成本，因而在过去几年，线上渠道的销量持续走高。尤其是直播和短视频电商的崛起，深刻影响了用户的消费行为，越来越多的读者通过短视频发掘书籍。对于商家而言，面对用户消费习惯和渠道的转变，如何顺应时代发展的趋势抓住新的增长机遇，成为现阶段的一大挑战。

(2) 阅读方式转变，数字阅读已具一定规模。

阅读方式方面，互联网的发展改变了读者的阅读方式。由于具有存储量大、检索便捷、便于保存、传播迅速等优势，数字阅读备受青睐，电子阅读与有声阅读市场已具备相当规模。中国音像与数字出版协会发布的《2021年度中国数字阅读报告》显示，数字阅读市场整体营收规模达415.7亿元，同比增长18.23%，其中大众阅读市场规模达302.5亿元，有声阅读达85.5亿元，专业阅读达27.7亿元；用户规模达5.06亿，其中44.63%为19~25岁用户，27.25%为18岁以下用户，年轻人成为数字阅读主力军。数字阅读将开启一个全新的阅读时代。对个人而言，书籍是不可或缺的精神食粮来源。书籍内容能与多样态的技术形式进行融合叠加，从而带给人们多感官、多时空、多方式的阅读体验。新的阅读模式也将影响未来社会的文化形态、社交模式等。

7.2 系统分析

本项目是一个数据分析项目，首先获取某出版社的图书信息，然后可视化分析这些图书信息。

7.2.1 系统介绍

××出版社成立于 1953 年 10 月，是工业和信息化部主管的大型专业出版社。自建社以来，它始终坚持正确的出版导向，坚持为科技发展与社会进步服务，为繁荣社会主义文化服务，坚持积极进取、改革创新，围绕"立足信息产业，面向现代社会，传播科学知识，服务科教兴国，为走中国特色新型工业化道路服务"的出版宗旨，已发展成为集图书、期刊、音像电子及数字出版于一体的综合性出版大社。

本项目爬虫模块的爬取目标是××出版社官网中的图书信息，在抓取到××出版社官网中的图书信息后，为了便于进行大数据分析，将抓取到的图书信息保存到 MySQL 数据库中，实现数据持久化处理。在大数据可视化分析模块中，使用 JSP 技术提取 MySQL 数据库中的图书信息，将大数据分析结果通过 Java Web 展示在网页前端，最终实现数据可视化分析功能。

7.2.2 需求分析

根据系统介绍，本系统需要完成以下两个方面的工作。

1) 网络爬虫

使用 Java 网络爬虫抓取××出版社官网中的图书信息，其中图书信息有多种主类别，例如计算机、电子、摄影、电影、音乐等。而在主类别下面又包含很多子类别，例如在"计算机"主类别下又包含办公软件、操作系统、移动开发、图形图像等子类别。本项目将编写独立的爬虫文件对特定类别的图书进行抓取，并进行单独保存。这样做的好处是采集到的数据比较清晰，使后面的大数据分析工作更加直观和易于理解。

2) 大数据分析

结合当前市场的热点和需求，利用大数据分析主要类别的图书信息。为了实现大数据分析的可视化功能，使用 Java Web 技术展示分析结果。为了使大数据分析结果更加直观，特意使用图表来展示分析结果。因为系统只是简单地提取并展示 MySQL 数据库中的数据，所以使用 JSP 技术实现前端展示功能，并没有采用 Struts、Spring 等专业级框架。

7.3 系统模块和实现流程

本项目各个模块的具体实现流程如图 7-1 所示。

第 7 章　图书市场数据分析系统

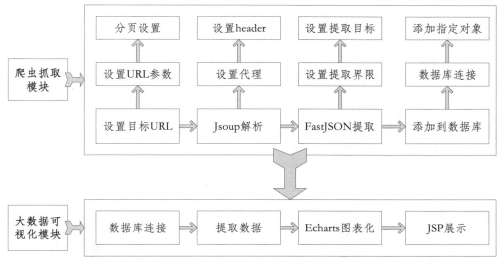

图 7-1　各个模块的具体实现流程

7.4　爬虫抓取模块

本节将详细讲解本项目中爬虫抓取模块的具体实现流程以及破解 post 方式反爬机制的方法。

扫码看视频

7.4.1　网页概览

××出版社"图书"主页的链接是 http://www.××.com.cn/shopping/index，在此页面会显示最热销的 9 种图书，如图 7-2 所示。

在畅销书上面显示主分类和子分类，其中主分类的界面效果如图 7-3 所示。

每个主分类下面会有多个子分类，例如，"计算机"主分类下的子分类如图 7-4 所示。

主分类"计算机"的 URL 地址是：

```
http://www.××.com.cn/shopping/search?tag=search&orderStr=hot&level1=2725fe7b-b2c2-4769-8f6f-c95f04c70275
```

如果单击主分类"摄影"，会发现此主分类的 URL 地址是：

```
http://www.××.com.cn/shopping/search?tag=search&orderStr=hot&level1=2725fe7b-b2c2-4769-8f6f-c95f04c70275
```

如果单击某个主分类下面的子分类，例如，单击主分类"计算机"下的"办公软件"，会发现此子分类的 URL 地址是：

```
http://www.xx.com.cn/shopping/search?tag=search&orderStr=hot&level1=2725fe7b-
b2c2-4769-8f6f-c95f04c70275
```

经过几次验证后会发现，所有的主分类 URL 地址和子分类 URL 地址相同，都是：

```
http://www.xx.com.cn/shopping/search?tag=search&orderStr=hot&level1=2725fe7b-
b2c2-4769-8f6f-c95f04c70275
```

并且此 URL 地址的网页通过分页展示了对应的图书信息，例如截止到目前(笔者编写本书时)主分类"计算机"有 429 个分页信息，如图 7-5 所示。

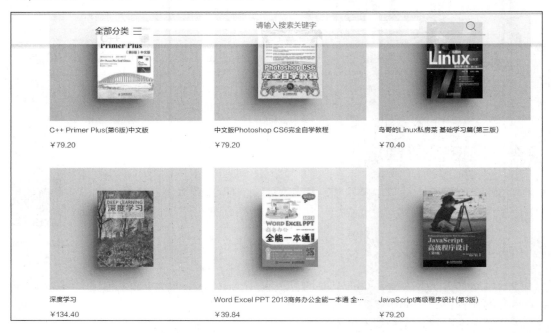

图 7-2 "图书"主页面

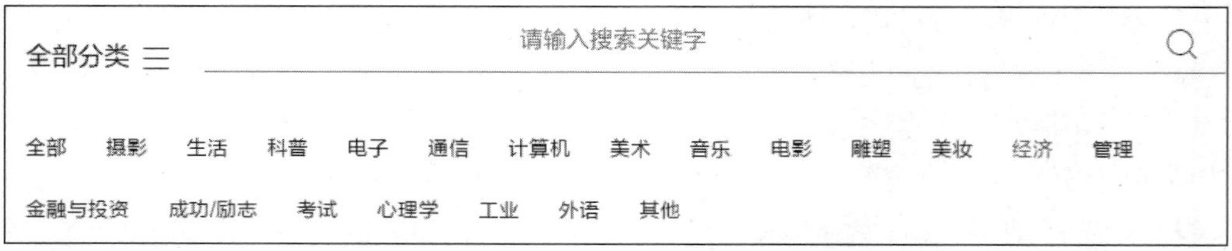

图 7-3 系统主分类

第 7 章 图书市场数据分析系统

图 7-4 "计算机"下面的子分类

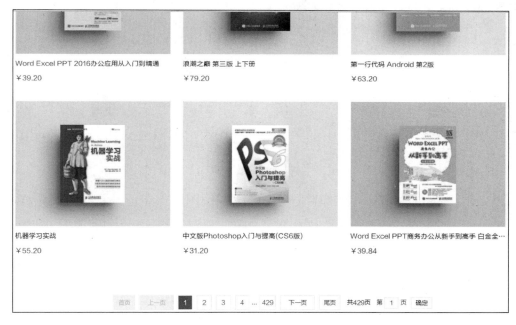

图 7-5 主分类"计算机"下的图书信息

如果在搜索表单中输入关键字并单击搜索按钮 🔍，会显示对应的搜索结果，例如，输入关键字"Python"后会搜索出所有的 Python 图书。此时搜索页面一共有以下两种和搜索功能相关的 URL 地址。

(1) 如果是在某分类 URL 地址中输入关键字进行检索，那么检索后的 URL 地址和分类地址相同，还是：

```
http://www.××.com.cn/shopping/search?tag=search&orderStr=hot&level1=2725fe7b-b2c2-4769-8f6f-c95f04c70275
```

(2) 如果是在图书主页 http://www.××.com.cn/shopping/index 中输入关键字进行检索，则

263

检索后的 URL 地址是：

http://www.xx.com.cn/shopping/search?tag=search&searchName=Python

上述 URL 地址中的"Python"表示输入的搜索关键字是 Python。

上述两种类型的搜索页面的展示效果相同，都是分页展示对应的搜索结果，例如，输入关键字"Java"后的搜索结果如图 7-6 所示。

图 7-6　搜索关键字为"Java"的图书

7.4.2　破解 JS API 反爬机制

经过前面的网页概览分析可知，××出版社的图书展示页面采用了 JavaScript 链接形式，所有分类的 URL 地址都是相同的。这给爬虫工作带来了困难，需要在开发模式中分析出真正的 URL 关系。例如，在搜索 Java 图书时显示的 URL 地址是：

http://www.xx.com.cn/shopping/search?tag=search&searchName=Java

按键盘上的 F12 键，进入浏览器的开发模式，单击 Network 按钮，此时图书信息展示页面和搜索页面的真正 URL 地址相同，如图 7-7 所示。

第 7 章 图书市场数据分析系统

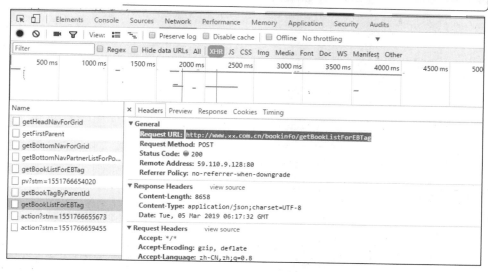

图 7-7 开发模式

也就是说，图书信息展示页面、分类图书展示页面和搜索页面的真正 URL 地址都是：

http://www.xx.com.cn/bookinfo/getBookListForEBTag

要想用爬虫提取图书信息，只需分析上述 URL 地址的传输内容即可。在 Headers 选项卡中会显示爬虫模块用到的参数，例如 User-Agent 代理参数和 URL 地址中的构成参数。Java 搜索页面的参数如图 7-8 所示。

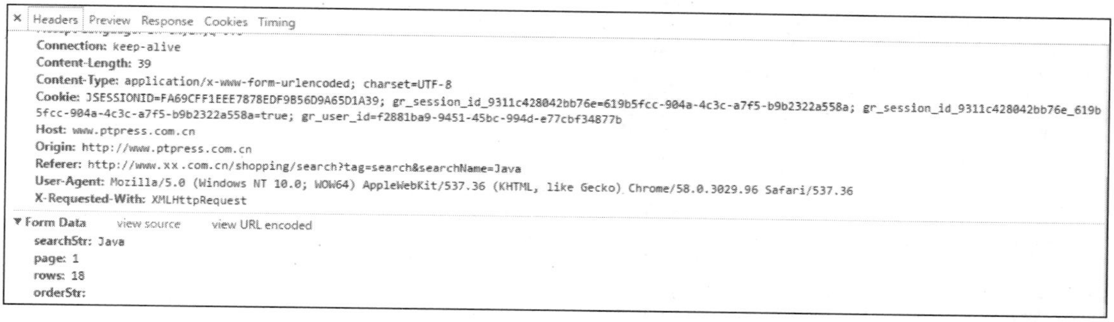

图 7-8 爬虫参数

在爬虫参数中，Form Data 选项卡中保存了和真正 URL 地址有关的构成参数，例如，Java 搜索页面的 URL 构成参数如下：

```
searchStr:Java
page:1
```

265

```
rows:18
orderStr:
```

其中，searchStr 表示搜索关键，page 表示分页序号，rows 表示每个分页显示的图书数量，orderStr 表示排序方式。

在右侧顶部单击 Response 选项，会显示获取到的图书数据，如图 7-9 所示。

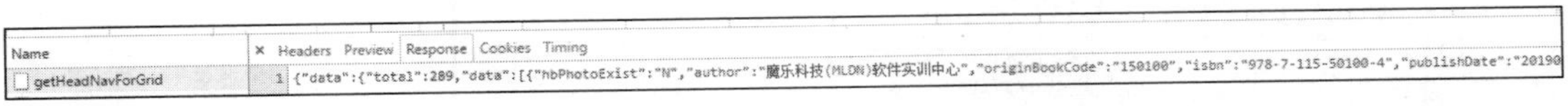

图 7-9　显示获取到的图书数据

单击 Response 选项后显示的图书数据是 JSON 格式的，例如，在 Java 搜索页面 URL 中获得的 JSON 数据如下。

```
{"data":{"total":289,"data":[{"hbPhotoExist":"N","author":"魔乐科技(MLDN)软件实训中心","originBookCode":"150100","isbn":"978-7-115-50100-4","publishDate":"201903","discountPrice":"95.20","bookDiscount":"0.8","bookName":"Java 编程技术大全(套装上下册)","executiveEditor":"张 翼","bookId":"22288273-61ff-4f25-8913-333c1baec84a","picPath":"http://47.93.163.221:8084/uploadimg/Material/978-7-115-50100-4/72jpg/50100_s300.jpg","price":"119","stockInDate":"20190304","shopType":"1","sumVolume":"0"},{"hbPhotoExist":"N","author":"张桓 李金靖","originBookCode":"150193","isbn":"978-7-115-50193-6","publishDate":"201902","discountPrice":"38.40","bookDiscount":"0.8","bookName":"Java Web 动态网站开发(微课版)","executiveEditor":"刘佳","bookId":"c14abdde-f5d9-4239-b12e-846c2c18b5a4","picPath":"http://47.93.163.221:8084/uploadimg/Material/978-7-115-50193-6/72jpg/50193_s300.jpg","price":"48","stockInDate":"20190213","shopType":"1","sumVolume":"0"},{"hbPhotoExist":"N","author":"田春瑾","originBookCode":"150420","isbn":"978-7-115-50420-3","publishDate":"201902","discountPrice":"7.24","bookDiscount":"0.8","bookName":"Java 程序设计习题与实践(微课版)","executiveEditor":"刘海溧","bookId":"1b13393b-db15-408d-aecf-5e6d8b8293aa","picPath":"http://47.93.163.221:8084/uploadimg/Material/978-7-115-50420-3/72jpg/50420_s300.jpg","price":"32.8","stockInDate":"20190128","shopType":"1","sumVolume":"0"},{"hbPhotoExist":"N","author":"普 运 伟","originBookCode":"150419","isbn":"978-7-115-50419-7","publishDate":"201902","discountPrice":"35.84","bookDiscount":"0.8","bookName":"Java 程序设计(微课版)","executiveEditor":"刘海溧","bookId":"d1ddd1ef-a005-4a88-8eec-4bc408642677","picPath":"http://47.93.163.221:8084/uploadimg/Material/978-7-115-50419-7/72jpg/50419_s300.jpg","price":"44.8","stockInDate":"20190128","shopType":"1","sumVolume":"0"},{"hbPhotoExist":"N","author":"赵建保","originBookCode":"149916","isbn":"978-7-115-49916-5","publishDate":"201902","discountPrice":"47.84","bookDiscount":"0.8","bookName":"JavaScript 前端开发模块化教程","executiveEditor":"范博涛","bookId":"daeb2e6c-6f3a-4d58-9147-491e96ead001","picPath":"http://47.93.163.221:8084/uploadimg/Material/978-7-115-49916-5/72jpg/49916_s300.jpg","price":"59.8","stockInDate":"20190128","shopType":"1","sumVolume":"0"},{"hbPhotoExist":"N","author":"刘彦君 张仁伟 满志强","originBookCode":"149179","isbn":"978-7-115-49179-4","publishDate":"201811","discountPrice":"55.84","bookDiscount":"0.8","bookName":"Java 面向对象思想与程序设计","executiveEditor":"税 梦 玲","bookId":"1ac50c73-5071-4309-8715-ff0ae742de1a","picPath":"http:
```

//47.93.163.221:8084/uploadimg/Material/978-7-115-49179-4/72jpg/49179_s300.jpg","price":"69.8","stockInDate":"20190115","shopType":"1","sumVolume":"0"},{"hbPhotoExist":"N","author":"潘俊","originBookCode":"149993","isbn":"978-7-115-49993-6","publishDate":"201901","discountPrice":"47.20","bookDiscount":"0.8","bookName":"JavaScript 函数式编程思想","executiveEditor":"张爽","bookId":"bfb30dc8-3e95-4036-9a83-fdd5802fe4d1","picPath":"http://47.93.163.221:8084/uploadimg/Material/978-7-115-49993-6/72jpg/49993_s300.jpg","price":"59","stockInDate":"20181229","shopType":"1","sumVolume":"0"},{"hbPhotoExist":"N","author":"戴雯惠 李家兵","originBookCode":"149749","isbn":"978-7-115-49749-9","publishDate":"201901","discountPrice":"39.84","bookDiscount":"0.8","bookName":"JavaScript+jQuery 开发实战","executiveEditor":"祝智敏","bookId":"9f9789cb-020f-4230-9d47-4ba1471fecf7","picPath":"http://47.93.163.221:8084/uploadimg/Material/978-7-115-49749-9/72jpg/49749_s300.jpg","price":"49.8","stockInDate":"20181213","shopType":"1","sumVolume":"0"},{"hbPhotoExist":"N","author":"吴以欣 陈小宁","originBookCode":"148775","isbn":"978-7-115-48775-9","publishDate":"201812","discountPrice":"39.84","bookDiscount":"0.8","bookName":"动态网页设计与制作(HTML5+CSS3+JavaScript)(第 3 版)","executiveEditor":"左仲海","bookId":"f56391e3-e2e3-4fb4-984d-1517a9aa6ddf","picPath":"http://47.93.163.221:8084/uploadimg/Material/978-7-115-48775-9/72jpg/48775_s300.jpg","price":"49.8","stockInDate":"20181205","shopType":"1","sumVolume":"403"},{"hbPhotoExist":"N","author":"[美]约翰·哈伯德(John R. Hubbard)","originBookCode":"149486","isbn":"978-7-115-49486-3","publishDate":"201812","discountPrice":"63.20","bookDiscount":"0.8","bookName":"Java 数据分析指南","executiveEditor":"胡俊英","bookId":"4d9f24ce-c6f8-4acb-84de-0b8d4f9a56fb","picPath":"http://47.93.163.221:8084/uploadimg/Material/978-7-115-49486-3/72jpg/49486_s300.jpg","price":"79","stockInDate":"20181128","shopType":"1","sumVolume":"937"},{"hbPhotoExist":"N","author":"李玉臣 臧金梅","originBookCode":"148977","isbn":"978-7-115-48977-7","publishDate":"201901","discountPrice":"36.00","bookDiscount":"0.8","bookName":"JavaScript 前端开发程序设计教程(微课版)","executiveEditor":"马小霞","bookId":"eba34ca7-f129-4676-949e-a842a54ccacc","picPath":"http://47.93.163.221:8084/uploadimg/Material/978-7-115-48977-7/72jpg/48977_s300.jpg","price":"45","stockInDate":"20180929","shopType":"1","sumVolume":"1071"},{"hbPhotoExist":"N","author":"戴远泉 李超 秦争艳","originBookCode":"148965","isbn":"978-7-115-48965-4","publishDate":"201901","discountPrice":"36.80","bookDiscount":"0.8","bookName":"Java 高级程序设计实战教程","executiveEditor":"桑珊","bookId":"857f10df-8356-481c-88ec-eebfef2e0b3a","picPath":"http://47.93.163.221:8084/uploadimg/Material/978-7-115-48965-4/72jpg/48965_s300.jpg","price":"46","stockInDate":"20180928","shopType":"1","sumVolume":"633"},{"hbPhotoExist":"N","author":"[西]哈维尔·费尔南德斯·冈萨雷斯","originBookCode":"149166","isbn":"978-7-115-49166-4","publishDate":"201810","discountPrice":"71.20","bookDiscount":"0.8","bookName":"精通 Java 并发编程 第 2 版","executiveEditor":"岳新欣","bookId":"b12b4e11-a9f9-4b77-bfa7-c7321e6ee892","picPath":"http://47.93.163.221:8084/uploadimg/Material/978-7-115-49166-4/72jpg/49166_s300.jpg","price":"89","stockInDate":"20180928","shopType":"1","sumVolume":"2369"},{"hbPhotoExist":"N","author":"夏帮贵 刘凡馨","originBookCode":"148693","isbn":"978-7-115-48693-6","publishDate":"201812","discountPrice":"39.84","bookDiscount":"0.8","bookName":"JavaScript+jQuery 前端开发基础教程(微课版)","executiveEditor":"左仲海","bookId":"98ae8dd5-2b01-4068-b899-fb1a07a9e479","picPath":"http://47.93.163.221:8084/uploadimg/Material/978-7-115-48693-6/72jpg/48693_s300.jpg","price":"49.8","stockInDate":"20180905","shopType":

```
"1","sumVolume":"1455"},{"hbPhotoExist":"Y","author":"张玉宏","originBookCode":
"148547","isbn":"978-7-115-48547-2","publishDate":"201902","discountPrice":"63.84",
"bookDiscount":"0.8","bookName":"Java 从入门到精通 精粹版","executiveEditor":"张翼",
"bookId":"64f9f0b6-402a-460f-89a4-688cf9897dc8","picPath":"http://47.93.163.221:
8084/uploadimg/Material/978-7-115-48547-2/72jpg/48547_s300.jpg","price":"79.8",
"stockInDate":"20180827","shopType":"1","sumVolume":"8250"},{"hbPhotoExist":"N",
"author":"[美]肯·寇森(Ken Kousen)","originBookCode":"148880","isbn":"978-7-115-
48880-0","publishDate":"201808","discountPrice":"55.20","bookDiscount":"0.8",
"bookName":"Java 攻略 Java 常见问题的简单解法","executiveEditor":"朱巍","bookId":
"a26777eb-2412-4769-9adb-45d966de68a4","picPath":"http://47.93.163.221:8084/
uploadimg/Material/978-7-115-48880-0/72jpg/48880_s300.jpg","price":"69",
"stockInDate":"20180820","shopType":"1","sumVolume":"1984"},{"hbPhotoExist":"N",
"author":"闫俊伢    耿 强","originBookCode":"148466","isbn":"978-7-115-48466-6",
"publishDate":"201807","discountPrice":"55.84","bookDiscount":"0.8","bookName":
"HTML5+CSS3+JavaScript+jQuery 程序设计基础教程(第 2 版)","executiveEditor":"邹文波",
"bookId":"dff9a63e-4ea1-456e-9e1b-50dcb59a800e","picPath":"http://47.93.163.221:
8084/uploadimg/Material/978-7-115-48466-6/72jpg/48466_s300.jpg","price":"69.8",
"stockInDate":"20180803","shopType":"1","sumVolume":"790"},{"hbPhotoExist":"N",
"author":"[美]朱莉·C·梅洛尼(Julie·C·Meloni)","originBookCode":"148349","isbn":
"978-7-115-48349-2","publishDate":"201808","discountPrice":"79.20","bookDiscount":
"0.8","bookName":"PHP MySQL 和 JavaScript 入门经典 第 6 版","executiveEditor":"陈冀康",
"bookId":"1940a30f-13b9-4a35-ad61-5aee253bf98e","picPath":"http://47.93.163.221:8084/
uploadimg/Material/978-7-115-48349-2/72jpg/48349_s300.jpg","price":"99","stockInDate":
"20180723","shopType":"1","sumVolume":"1754"}]},"msg":"调用接口数据成功！","success":
true}
```

在上述 JSON 文件中，开始的"total":289 表示××出版社中所有 Java 相关图书数据有 289 条。后面的每一个 JSON 数据对应一种 Java 书的信息，每一种书的信息包含作者、ISBN、价格和图书图片等。因此，我们爬虫的目标就是确定要抓取的目标 URL，然后提取这个 URL 下面对应的 JSON 数据，最后将 JSON 中提取的图书信息添加到数据库中。

7.4.3 爬虫抓取 Java 图书信息

根据前面的破解 JS API 反爬机制内容可知，检索关键字"Java"后，显示 Java 相关图书的基本 URL 地址是 http://www.××.com.cn/bookinfo/getBookListForEBTag，后面的 URL 构成参数是 searchStr:Java 和页码数字。

编写文件 JavaBooK05.java，功能是抓取关键字为"Java"的所有图书信息，并将抓取的信息添加到 MySQL 数据库中。文件 JavaBooK05.java 的具体实现流程如下。

(1) 编写函数 mysqlinsert()，功能是使用 INSERT INTO 语句向指定的 MySQL 数据库添加信息，具体实现代码如下。

```java
public class PythonBook {
```

```java
//数据库插入信息
    public static void mysqlinsert(String author,String bookName, String price,
String bookId, String picPath, String data, String bookDiscount) {
        final String DB_URL = "jdbc:mysql://localhost:3306/chubanshe?useSSL=false";
        final String USER = "root";
        final String PASS = "66688888";

        Connection conn = null;
        Statement stmt = null;
        try{
            Class.forName("com.mysql.jdbc.Driver");
            conn = DriverManager.getConnection(DB_URL,USER,PASS);
            stmt = conn.createStatement();
            String sql;
            sql = "INSERT INTO pythonbooks (author,bookName, price, bookId,
picPath,data, bookDiscount) VALUES ('"+author+"','"+bookName+"','"+price+"',
'"+bookId+"','"+picPath+"','"+data+"','"+bookDiscount+"');";
            stmt.executeUpdate(sql);

            stmt.close();
            conn.close();
        }catch(SQLException se){
            se.printStackTrace();
        }catch(Exception e){
            e.printStackTrace();
        }finally{
            try{
                if(stmt!=null) stmt.close();
            }catch(SQLException se2){
            }
            try{
                if(conn!=null) conn.close();
            }catch(SQLException se){
                se.printStackTrace();
            }
        }
        return;
    }
```

(2) 设置要抓取的 URL 链接，在 Map 集合对象 m 中设置 URL 构成参数，通过 for 循环设置抓取 18 个分页，按 F12 键进入开发模式，分别设置参数 userAgent 和 header。具体实现代码如下。

```java
public static void main(String[] args) throws Exception {
    String bbb ="http://www.ptpress.com.cn/bookinfo/getBookListForEBTag";

    Map<String, String> m = new IdentityHashMap<String, String>();
```

```
            m.put("searchStr", "Python");
            m.put("rows", "18");
            for (int ddd = 0; ddd < 9; ddd++) {
                m.put("page", ddd + "");
                String body = Jsoup
                        .connect(bbb)
                        .ignoreContentType(true)
                        .ignoreHttpErrors(true)
                        .timeout(1000 * 30)
                        .userAgent("Mozilla/5.0 (Macintosh; Intel Mac OS X 10_13_3) AppleWebKit/537.36 (KHTML, like Gecko) Chrome/65.0.3325.181 Safari/537.36")
                        .header("accept","text/html,application/xhtml+xml,application/xml;q=0.9,image/webp,image/apng,*/*;q=0.8")
                        .header("accept-encoding","gzip, deflate, br")
                        .header("accept-language","zh-CN,zh;q=0.9,en-US;q=0.8,en;q=0.7")
                        .data(m)
                        .execute().body();
```

(3) 解析抓取到的 JSON 数据,然后提取指定的 JSON 对象添加到数据库中,具体实现代码如下。

```
                JSONObject jsonObject = JSONObject.parseObject(body);
                jsonObject.getJSONObject("data").getJSONArray("data").forEach(i -> {
                    String bookName = ((JSONObject)i).getString("bookName");
                    String author = ((JSONObject)i).getString("author");
                    String price = ((JSONObject)i).getString("price");
                    String bookId = ((JSONObject)i).getString("bookId");
                    String data = ((JSONObject)i).getString("stockInDate");
                    String picPath = ((JSONObject)i).getString("picPath");
                    String bookDiscount = ((JSONObject)i).getString("bookDiscount");
                    System.out.println(author + "-" + price + "-" + bookId + "-" + picPath
                            + "-" + bookDiscount);
                    mysqlinsert(bookName,author,price,bookId,picPath,data,bookDiscount);
                });
            }
        }
    }
```

在运行上述 Java 文件之前,需要先使用以下 SQL 语句在 MySQL 数据库中创建数据表 javabooks,用于保存抓取到的 Java 图书信息。

```
CREATE TABLE javabooks(
    id INT NOT NULL AUTO_INCREMENT,
    bookName VARCHAR(400) NOT NULL,
    price varchar(50) NOT NULL,
    bookId VARCHAR(400) NOT NULL,
    picPath VARCHAR(500) NOT NULL,
```

```
  author VARCHAR(400) NOT NULL,
  data VARCHAR(500) NOT NULL DEFAULT '',
  bookDiscount VARCHAR(500) NOT NULL,
  PRIMARY KEY (id)
)ENGINE=InnoDB DEFAULT CHARSET=utf8;
```

执行文件后，会在 Eclipse 控制台打印输出抓取到的信息，如图 7-10 所示。

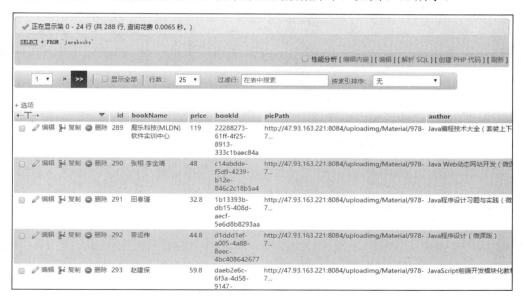

图 7-10　抓取到的信息

文件执行完毕，会将所有的 Java 图书添加到数据库中，如图 7-11 所示。

图 7-11　添加到数据库中的 Java 图书数据

7.4.4 爬虫抓取 Python 图书信息

编写程序文件 PythonBook.java，功能是抓取关键字为"Python"的所有图书信息，并将抓取的信息添加到 MySQL 数据库中。文件 PythonBook.java 和 JavaBooK05.java 类似，具体实现代码如下。

```java
public class PythonBook {
    //数据库插入信息
    public static void mysqlinsert(String author,String bookName, String price,
String bookId, String picPath, String data, String bookDiscount) {
        final String DB_URL = "jdbc:mysql://localhost:3306/chubanshe?useSSL=false";
        final String USER = "root";
        final String PASS = "66688888";

        Connection conn = null;
        Statement stmt = null;
        try{
            Class.forName("com.mysql.jdbc.Driver");
            conn = DriverManager.getConnection(DB_URL,USER,PASS);
            stmt = conn.createStatement();
            String sql;
            sql = "INSERT INTO pythonbooks (author,bookName, price, bookId,
picPath,data, bookDiscount) VALUES ('"+author+"','"+bookName+"','"+price+"',
'"+bookId+"','"+picPath+"','"+data+"','"+bookDiscount+"');";
            stmt.executeUpdate(sql);

            stmt.close();
            conn.close();
        }catch(SQLException se){
            se.printStackTrace();
        }catch(Exception e){
            e.printStackTrace();
        }finally{
            try{
                if(stmt!=null) stmt.close();
            }catch(SQLException se2){
            }
            try{
                if(conn!=null) conn.close();
            }catch(SQLException se){
                se.printStackTrace();
            }
        }
        return;
```

```java
    }
    public static void main(String[] args) throws Exception {
        String bbb ="http://www.xx.com.cn/bookinfo/getBookListForEBTag";

        Map<String, String> m = new IdentityHashMap<String, String>();
        m.put("searchStr", "Python");
        m.put("rows", "18");
        for (int ddd = 0; ddd < 9; ddd++) {
            m.put("page", ddd + "");
            String body = Jsoup
                    .connect(bbb)
                    .ignoreContentType(true)
                    .ignoreHttpErrors(true)
                    .timeout(1000 * 30)
                    .userAgent("Mozilla/5.0 (Macintosh; Intel Mac OS X 10_13_3) AppleWebKit/537.36 (KHTML, like Gecko) Chrome/65.0.3325.181 Safari/537.36")
                    .header("accept","text/html,application/xhtml+xml,application/xml;q=0.9,image/webp,image/apng,*/*;q=0.8")
                    .header("accept-encoding","gzip, deflate, br")
                    .header("accept-language","zh-CN,zh;q=0.9,en-US;q=0.8,en;q=0.7")
                    .data(m)
                    .execute().body();
//          System.out.println(body);
            JSONObject jsonObject = JSONObject.parseObject(body);
            jsonObject.getJSONObject("data").getJSONArray("data").forEach(i -> {
                String bookName = ((JSONObject)i).getString("bookName");
                String author = ((JSONObject)i).getString("author");
                String price = ((JSONObject)i).getString("price");
                String bookId = ((JSONObject)i).getString("bookId");
                String data = ((JSONObject)i).getString("stockInDate");
                String picPath = ((JSONObject)i).getString("picPath");
                String bookDiscount = ((JSONObject)i).getString("bookDiscount");
                System.out.println(author + "-" + price + "-" + bookId + "-" + picPath
                        + "-" + bookDiscount);
                mysqlinsert(bookName,author,price,bookId,picPath,data,bookDiscount);
            });
        }
    }
}
```

在运行上述 Java 文件之前,需要先使用以下 SQL 语句在 MySQL 数据库中创建数据表 pythonbooks,用于保存抓取到的 Python 图书信息。

```sql
CREATE TABLE pythonbooks(
    id INT NOT NULL AUTO_INCREMENT,
    bookName VARCHAR(400) NOT NULL,
    price varchar(50) NOT NULL,
```

```
  bookId VARCHAR(400) NOT NULL,
  picPath VARCHAR(500) NOT NULL,
  author VARCHAR(400) NOT NULL,
  data VARCHAR(500) NOT NULL DEFAULT '',
  bookDiscount VARCHAR(500) NOT NULL,
  PRIMARY KEY (id)
)ENGINE=InnoDB DEFAULT CHARSET=utf8;
```

文件执行完毕，会将所有的 Python 图书添加到数据库中，如图 7-12 所示。

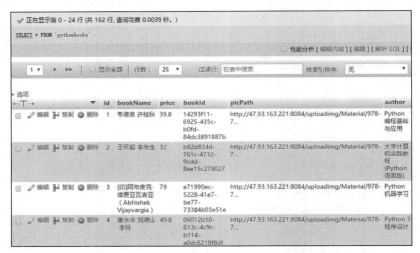

图 7-12　添加到数据库中的 Python 图书数据

7.4.5　爬虫抓取主分类图书信息类

(1) 假如想抓取主分类"计算机"下的所有图书信息，单击导航中的"计算机"链接，来到主分类页面，其中一共有 429 个分页，每个分页显示 18 本图书信息，如图 7-13 所示。

(2) 按 F12 键进入开发模式，单击 Network 按钮查看实际 URL，如图 7-14 所示。

在图 7-14 中可以看出，主类"计算机"的实际 URL 是 http://www.××.com.cn/bookinfo/getBookListForEBTag，构成的 URL 参数如下。

- page:1：表示当前分页数字。
- rows:18：表示每个分页显示 18 本图书信息。
- bookTagId:2725fe7b-b2c2-4769-8f6f-c95f04c70275：表示当前主类"计算机"对应的编号。
- orderStr:hot：表示排序顺序。

第 7 章　图书市场数据分析系统

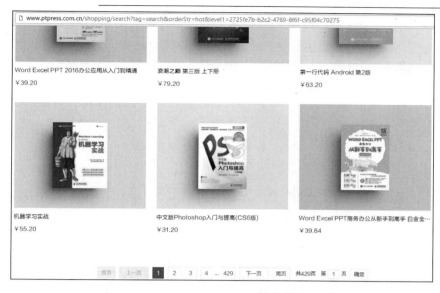

图 7-13　主类"计算机"下的所有图书信息列表

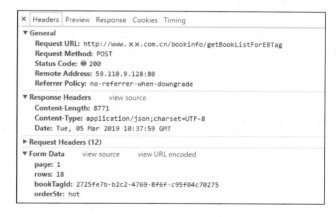

图 7-14　开发模式

从 Response 选项卡获取的 JSON 结果中可以看出，当前主分类"计算机"下的所有的图书信息共有 7715 本，如图 7-15 所示。

图 7-15　JSON 结果

(3) 编写程序文件 JSJBooks.java，功能是抓取主分类"计算机"下的所有图书信息，并将抓取的数据添加到系统数据库中。文件 JSJBooks.java 的主要实现代码如下。

```java
    public static void main(String[] args) throws Exception {
        String bbb ="http://www.××.com.cn/bookinfo/getBookListForEBTag";

        Map<String, String> m = new IdentityHashMap<String, String>();
        m.put("bookTagId", "2725fe7b-b2c2-4769-8f6f-c95f04c70275");
        m.put("rows", "18");
        m.put("orderStr", "hot");
        for (int ddd = 0; ddd < 429; ddd++) {
            m.put("page", ddd + "");
            String body = Jsoup
                    .connect(bbb)
                    .ignoreContentType(true)
                    .ignoreHttpErrors(true)
                    .timeout(1000 * 30)
                    .userAgent("Mozilla/5.0 (Macintosh; Intel Mac OS X 10_13_3) AppleWebKit/537.36 (KHTML, like Gecko) Chrome/65.0.3325.181 Safari/537.36")
                    .header("accept","text/html,application/xhtml+xml,application/xml;q=0.9,image/webp,image/apng,*/*;q=0.8")
                    .header("accept-encoding","gzip, deflate, br")
                    .header("accept-language","zh-CN,zh;q=0.9,en-US;q=0.8,en;q=0.7")
                    .data(m)
                    .execute().body();
//            System.out.println(body);
            JSONObject jsonObject = JSONObject.parseObject(body);
            jsonObject.getJSONObject("data").getJSONArray("data").forEach(i -> {
                String bookName = ((JSONObject)i).getString("bookName");
                String author = ((JSONObject)i).getString("author");
                String price = ((JSONObject)i).getString("price");
                String bookId = ((JSONObject)i).getString("bookId");
                String picPath = ((JSONObject)i).getString("picPath");
                String bookDiscount = ((JSONObject)i).getString("bookDiscount");
                System.out.println(author + "-" + price + "-" + bookId + "-" + picPath
                        + "-" + bookDiscount);
                mysqlinsert(bookName,author,price,bookId,picPath,bookDiscount);
            });
        }
    }
```

执行文件,将抓取到的计算机图书信息添加到数据库中,如图 7-16 所示。

(4) 用同样的方法,抓取其他主分类下的图书信息。假如想抓取主分类"经济"下的所有图书信息,单击导航中的"经济"链接来到主分类页面,其中一共有 81 个分页,每个分页显示 18 本图书信息,如图 7-17 所示。

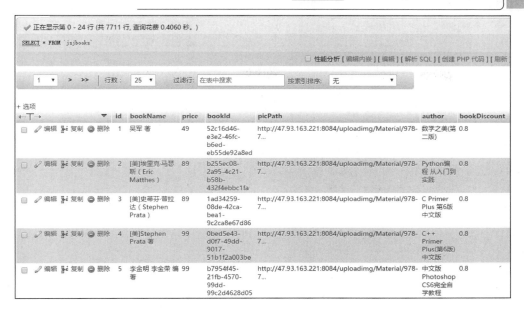

图 7-16 添加到数据库中的"计算机"图书数据

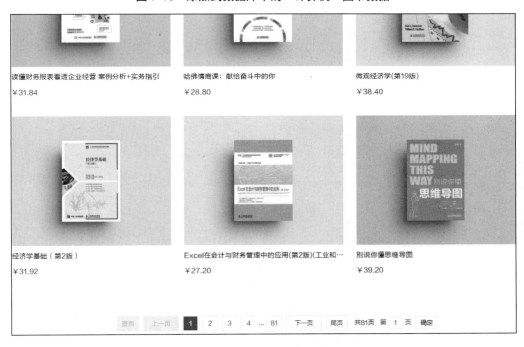

图 7-17 主分类"经济"下的所有图书信息列表

(5) 按 F12 键进入开发模式，单击 Network 按钮查看实际 URL，如图 7-18 所示。

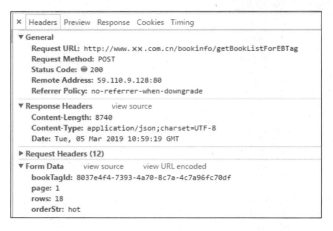

图 7-18　开发模式

在图 7-18 中可以看出，主类"计算机"的实际 URL 是 http://www.××.com.cn/bookinfo/getBookListForEBTag，构成的 URL 参数如下。

- page:1：表示当前分页数字。
- rows:18：表示每个分页显示 18 本图书信息。
- bookTagId:8037e4f4-7393-4a70-8c7a-4c7a96fc70df：表示当前主类"经济"对应的编号。
- orderStr:hot：表示排序顺序。

从 Response 选项卡获取的 JSON 结果中可以看出，当前主分类"经济"下的所有的图书信息共有 1442 本，如图 7-19 所示。

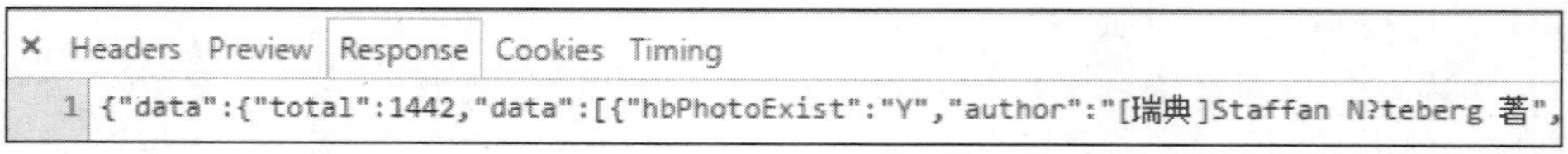

图 7-19　JSON 结果

(6) 编写程序文件 JingjiBook.java，功能是抓取主分类"经济"下的所有图书信息，并将抓取的数据添加到系统数据库中，实现代码如下。

```java
public class JingjiBook {
    //数据库插入信息
    public static void mysqlinsert(String author,String bookName, String price, String bookId, String picPath, String bookDiscount) {
        final String DB_URL = "jdbc:mysql://localhost:3306/chubanshe?useSSL=false";
```

```java
        final String USER = "root";
        final String PASS = "66688888";

        Connection conn = null;
        Statement stmt = null;
        try{
            Class.forName("com.mysql.jdbc.Driver");
            conn = DriverManager.getConnection(DB_URL,USER,PASS);
            stmt = conn.createStatement();
            String sql;
            sql = "INSERT INTO JSJBooks (author,bookName, price, bookId, picPath, bookDiscount) VALUES ('"+author+"','"+bookName+"','"+price+"','"+bookId+"','"+picPath+"','"+bookDiscount+"');";
            stmt.executeUpdate(sql);

            stmt.close();
            conn.close();
        }catch(SQLException se){
            se.printStackTrace();
        }catch(Exception e){
            e.printStackTrace();
        }finally{
            try{
                if(stmt!=null) stmt.close();
            }catch(SQLException se2){
            }
            try{
                if(conn!=null) conn.close();
            }catch(SQLException se){
                se.printStackTrace();
            }
        }
        return;
    }
    public static void main(String[] args) throws Exception {
        String bbb ="http://www.ptpress.com.cn/bookinfo/getBookListForEBTag";

        Map<String, String> m = new IdentityHashMap<String, String>();
        m.put("bookTagId", "8037e4f4-7393-4a70-8c7a-4c7a96fc70df");
        m.put("rows", "18");
        m.put("orderStr", "hot");
        for (int ddd = 0; ddd < 80; ddd++) {
            m.put("page", ddd + "");
            String body = Jsoup
                    .connect(bbb)
                    .ignoreContentType(true)
                    .ignoreHttpErrors(true)
                    .timeout(1000 * 30)
```

```
                    .userAgent("Mozilla/5.0 (Macintosh; Intel Mac OS X 10_13_3)
AppleWebKit/537.36 (KHTML, like Gecko) Chrome/65.0.3325.181 Safari/537.36")
                    .header("accept","text/html,application/xhtml+xml,application/
                            xml;q=0.9,image/webp,image/apng,*/*;q=0.8")
                    .header("accept-encoding","gzip, deflate, br")
                    .header("accept-language","zh-CN,zh;q=0.9,en-US;q=0.8,en;q=0.7")
                    .data(m)
                    .execute().body();
//          System.out.println(body);
            JSONObject jsonObject = JSONObject.parseObject(body);
            jsonObject.getJSONObject("data").getJSONArray("data").forEach(i -> {
                String bookName = ((JSONObject)i).getString("bookName");
                String author = ((JSONObject)i).getString("author");
                String price = ((JSONObject)i).getString("price");
                String bookId = ((JSONObject)i).getString("bookId");
                String picPath = ((JSONObject)i).getString("picPath");
                String bookDiscount = ((JSONObject)i).getString("bookDiscount");
                System.out.println(author + "-" + price + "-" + bookId + "-" + picPath
                        + "-" + bookDiscount);
                mysqlinsert(bookName,author,price,bookId,picPath,bookDiscount);
            });
        }
    }
}
```

(7) 用同样的方法，编写对应的 Java 程序文件，依次抓取以下主类下的所有图书信息。

❑ 文件 kepuBook.java：抓取主类"科普"下的所有图书信息。

❑ 文件 guanliBook.java：抓取主类"管理"下的所有图书信息。

❑ 文件 sheyingBook.java：抓取主类"摄影"下的所有图书信息。

❑ 文件 shenghuoBook.java：抓取主类"生活"下的所有图书信息。

❑ 文件 jingjiBook.java：抓取主类"经济"下的所有图书信息。

❑ 文件 dianziBook.java：抓取主类"电子"下的所有图书信息。

7.4.6 爬虫抓取子分类图书信息类

(1) 假如想抓取主分类"计算机"下子分类"多媒体"的所有图书信息，单击导航中的"多媒体"链接来到子分类页面，其中一共有 28 个分页，每个分页显示 18 本图书信息，如图 7-20 所示。

(2) 按 F12 键进入开发模式，单击 Network 按钮查看实际 URL，如图 7-21 所示。

在图 7-21 中可以看出，子分类"多媒体"的实际 URL 是 http://www.××.com.cn/bookinfo/getBookListForEBTag，构成的 URL 参数如下。

- page:1：表示当前分页数字。
- rows:18：表示每个分页显示18本图书信息。
- bookTagId: e534408a-d8f5-45e8-bde7-a832d09c03f5：表示当前子分类"多媒体"对应的编号。
- orderStr:hot：表示排序顺序。

图 7-20　主分类"计算机"下子分类"多媒体"的所有图书信息

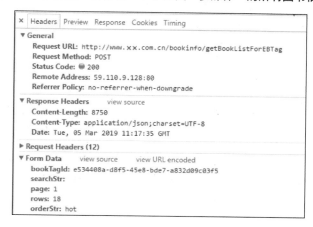

图 7-21　开发模式

从 Response 选项卡获取的 JSON 结果中可以看出,当前子分类"多媒体"下的所有的图书信息共有 504 本,如图 7-22 所示。

| Headers | Preview | Response | Cookies | Timing |

1 {"data":{"total":504,"data":[{"hbPhotoExist":"N","author":"新视角文化行 编著","originBookCode":"130092",

图 7-22 JSON 结果

(3) 编写程序文件 JSJDuomeitiBooks,功能是抓取主分类"计算机"下子分类"多媒体"的所有图书信息,并将抓取的数据添加到系统数据库中。文件 JSJDuomeitiBooks 的主要实现代码如下。

```java
public static void main(String[] args) throws Exception {
    String bbb ="http://www.xx.com.cn/bookinfo/getBookListForEBTag";

    Map<String, String> m = new IdentityHashMap<String, String>();
    m.put("bookTagId", "e534408a-d8f5-45e8-bde7-a832d09c03f5");
    m.put("rows", "18");
    m.put("orderStr", "hot");
    for (int ddd = 0; ddd < 28; ddd++) {
        m.put("page", ddd + "");
        String body = Jsoup
                .connect(bbb)
                .ignoreContentType(true)
                .ignoreHttpErrors(true)
                .timeout(1000 * 30)
                .userAgent("Mozilla/5.0 (Macintosh; Intel Mac OS X 10_13_3) AppleWebKit/537.36 (KHTML, like Gecko) Chrome/65.0.3325.181 Safari/537.36")
                .header("accept","text/html,application/xhtml+xml,application/xml;q=0.9,image/webp,image/apng,*/*;q=0.8")
                .header("accept-encoding","gzip, deflate, br")
                .header("accept-language","zh-CN,zh;q=0.9,en-US;q=0.8,en;q=0.7")
                .data(m)
                .execute().body();
//            System.out.println(body);
        JSONObject jsonObject = JSONObject.parseObject(body);
        jsonObject.getJSONObject("data").getJSONArray("data").forEach(i -> {
            String bookName = ((JSONObject)i).getString("bookName");
            String author = ((JSONObject)i).getString("author");
            String price = ((JSONObject)i).getString("price");
            String bookId = ((JSONObject)i).getString("bookId");
            String picPath = ((JSONObject)i).getString("picPath");
            String bookDiscount = ((JSONObject)i).getString("bookDiscount");
```

```
            System.out.println(author + "-" + price + "-" + bookId + "-" + picPath
                    + "-" + bookDiscount);
            mysqlinsert(bookName,author,price,bookId,picPath,bookDiscount);
        });
    }
}
```

执行上述代码后，会将抓取到的数据添加到系统数据库中，如图7-23所示。

id	bookName	price	bookId	picPath	author	bookDiscount
1	新视角文化行编著	49.8	adcfca34-12a4-4357-aa1f-bfa083e2d766	http://47.93.163.221:8084/uploadimg/Material/978-7...	Flash CS6动画制作实战从入门到精通	0.8
2	[英]马克西姆·亚戈（Maxim Jago）	79	93281b40-c164-4767-94aa-901e41da8787	http://47.93.163.221:8084/uploadimg/Material/978-7...	Adobe Premiere Pro CC 2017经典教程	0.8
3	[美]Adobe公司	59	31c51048-2cca-4d08-9fb6-2af694ed59eb	http://47.93.163.221:8084/uploadimg/Material/978-7...	Adobe Premiere Pro CC经典教程	0.8
4	[美]Adobe公司 著	49	b1d5c3d6-dba0-4647-807a-7532bd1c7aff	http://47.93.163.221:8084/uploadimg/Material/978-7...	Adobe After Effects CC经典教程	0.8
5	程明才 编著	99	3a0d328d-ede7-4447-948d-6d21c05929ff	http://47.93.163.221:8084/uploadimg/Material/978-7...	After Effects CC中文版超级学习手册	0.8

图7-23 抓取到的图书数据被保存在数据库中

(4) 用同样的方法编写对应的Java程序文件，依次抓取主分类"计算机"下子分类的所有图书信息。

- 文件JSJKaoshiBooks.java：抓取子分类"考试"下的所有图书信息。
- 文件JSJMobileBooks.java：抓取子分类"移动"下的所有图书信息。
- 文件JSJOfficeBooks.java：抓取子分类"办公"下的所有图书信息。
- 文件JSJShejiBooks.java：抓取子分类"设计"下的所有图书信息。
- 文件JSJTuxingBook.java：抓取子分类"图形图像"下的所有图书信息。
- 文件JSJTongxinBooks.java：抓取子分类"通信"下的所有图书信息。

7.5 大数据可视化分析

上一节已经将主要的图书信息爬取完毕，接下来将抓取到的数据进行可视化分析，实现大数据分析和提取工作，这样可以将数据更好地在工作和生活中使用。

扫码看视频

7.5.1 搭建 Java Web 平台

Tomcat 是 Java Web 运行的服务器软件，用户要开发和运行 Java Web 程序，就必须对它进行下载和安装。在安装 Tomcat 之前，一定要安装和配置 JDK。

（1）打开浏览器，在地址栏中输入 http://tomcat.apache.org/进行浏览，如图 7-24 所示。

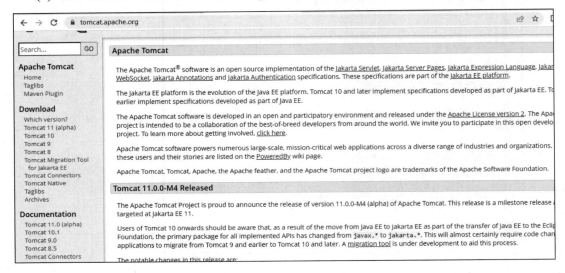

图 7-24 Tomcat 的首页

（2）单击左侧的 Tomcat 7.x 超级链接，在打开的新页面中，将网页滑动到最下面，如图 7-25 所示。

（3）单击 32-bit/64-bit Windows Service Installer (pgp, sha512)链接，等待完成下载。

（4）双击下载的 Tomcat 软件，即可对它进行安装，单击 Next 按钮，在安装过程中建议使用默认设置。

（5）设置安装目录，例如 "H:\jsp"。

（6）设置服务器的端口，Tomcat 的默认端口是 8080，我们可以将其修改为自己喜欢的，例如 8089，然后设置管理用户名的密码，如图 7-26 所示。

（7）在打开的窗口中选择 JDK 里的 JRE 文件，在默认情况下，它会寻找到这个文件，如果寻找不到，用户需要对它进行设置。设置好后单击 Install 按钮进行安装，耐心等待安装完成即可。

（8）安装完成，在任务栏中双击 Tomcat 图标，然后单击 Start 按钮启动 Tomcat 服务器，如图 7-27 所示。

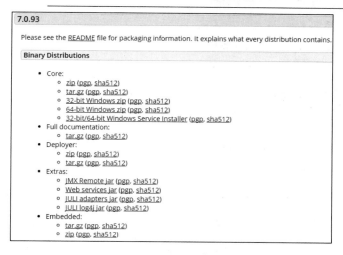

图 7-25　下载最新版本

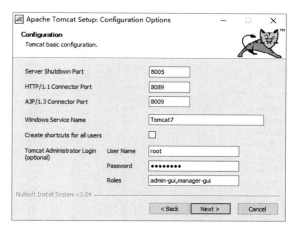

图 7-26　设置 Tomcat 服务器

图 7-27　启动 Tomcat 服务器

　　(9) 在浏览器中输入测试地址即可显示 Tomcat 服务器主页，测试地址是 http://127.0.0.1:8089/，其中 8089 是在前面设置的 HTTP 端口号，如图 7-28 所示。

　　(10) 因为前文设置的安装目录是"H:\jsp"，打开此目录后会显示安装的服务器文件，在 webapps 文件夹中保存了 Java Web 程序文件，如图 7-29 所示。

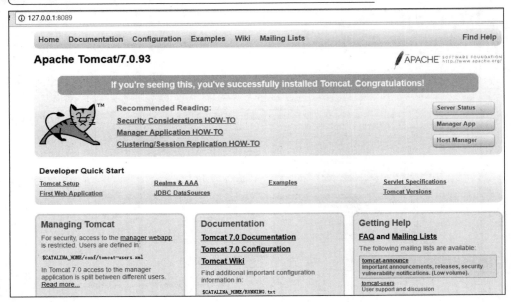

图 7-28　Tomcat 服务器主页

图 7-29　显示安装的服务器文件

7.5.2　大数据分析并可视化计算机图书数据

（1）在服务器目录"H:\jsp\webapps\"中新建文件夹 26，用于保存本项目所需要的 Java Web 程序。

（2）将本项目中需要的 jar 库文件复制到目录 H:\jsp\lib 中，如图 7-30 所示。其中有些是库文件。

图 7-30 需要用到的 jar 库文件

(3) 编写程序文件 JSJBar.jsp，功能是分别提取数据库中的 Java、Python、多媒体、考试、移动开发、办公和辅助设计等类型图书的数据，然后通过柱状图实现可视化分析。文件 JSJBar.jsp 的具体实现流程如下。

① 引入 JSP 指令和数据库连接需要的头文件，具体实现代码如下。

```
<%@ page language="java" contentType="text/html; charset=UTF-8"
    pageEncoding="UTF-8"%>
<%@ page import="java.io.*,java.util.*,java.sql.*"%>
<%@ page import="javax.servlet.http.*,javax.servlet.*" %>
<%@ taglib uri="http://java.sun.com/jsp/jstl/core" prefix="c"%>
<%@ taglib uri="http://java.sun.com/jsp/jstl/sql" prefix="sql"%>
```

② 引入 Echarts 头文件，功能是实现图表可视化效果，具体实现代码如下。

```
<html>
<head>
<!-- 引入 echarts.js -->
<script src="echarts.common.min.js"></script>
<title>计算机图书统计柱状图</title>
</head>
<body>
```

③ 使用 JDBC 建立和数据库的连接，然后使用 SQL 语句查询并统计 Java、Python、多

媒体、考试、移动开发、办公和辅助设计等数据库表中的数据条数,具体实现代码如下。

```xml
<!--
JDBC 驱动名及数据库 URL
数据库的用户名与密码,根据自己的需要设置
useUnicode=true&characterEncoding=utf-8 防止中文乱码
 -->
<sql:setDataSource var="snapshot" driver="com.mysql.jdbc.Driver"
    url="jdbc:mysql://localhost:3306/chubanshe?&characterEncoding=utf-8"
    user="root"  password="66688888"/>

<sql:query dataSource="${snapshot}" var="result">
select t1.num1,t2.num2,t3.num3,t4.num4,t5.num5,t6.num6,t7.num7 from (select
count(bookName) num1 from javabooks) t1, (select count(bookName) num2 from
pythonbooks) t2, (select count(bookName) num3 from jsjduomeitibooks) t3, (select
count(bookName) num4 from jsjkaoshibooks) t4, (select count(bookName) num5 from
jsjmobilebooks) t5, (select count(bookName) num6 from jsjofficebooks) t6, (select
count(bookName) num7 from jsjshejibooks) t7;
</sql:query>
```

④ 将统计的各类图书数据放在 Echarts 图表中作为可视化素材数据,具体实现代码如下。

```html
<c:forEach var="row" items="${result.rows}">

   <div id="main" style="width: 600px;height:400px;"></div>
   <script type="text/javascript">
       // 基于准备好的dom,初始化Echarts实例
       var myChart = echarts.init(document.getElementById('main'));

       // 指定图表的配置项和数据
       var option = {
           title: {
               text: '××出版社计算机图书主要类别的作品数量统计'
           },
           tooltip: {},

           xAxis: {
               data: ["Java","Python","多媒体","考试","移动开发","办公","辅助设计"]
           },
           yAxis: {},
           series: [{
               name: '此类书数量',
               type: 'bar',
               data: [<c:out value="${row.num1}"/>, <c:out value="${row.num2}"/>,
<c:out value="${row.num3}"/>, <c:out value="${row.num4}"/>, <c:out
value="${row.num5}"/>, <c:out value="${row.num6}"/>,<c:out value="${row.num7}"/>]
```

```
                  }]
            };
                // 使用刚指定的配置项和数据显示图表
            myChart.setOption(option);
    </script>
      </c:forEach>
</body>
</html>
```

在浏览器中输入 http://127.0.0.1:8089/26/JSJBar.jsp，会显示计算机图书统计柱状图效果，如图 7-31 所示。其中，8089 是端口号，26 是在 Tomcat 服务器中设置的文件夹。

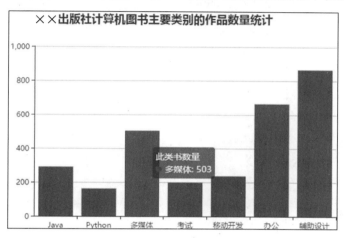

图 7-31　计算机图书数据可视化

7.5.3　大数据分析并可视化近期 Java 书和 Python 书的数据

本节将编写 Java Web 程序文件 JavaPythondui.jsp，大数据分析并可视化 2023 年现有 Java 书和 Python 书的数据。文件 JavaPythondui.jsp 的主要实现代码如下。

```
<!-- 引入 echarts.js -->
<script src="echarts.common.min.js"></script>
<title>计算机图书统计柱状图</title>
</head>
<body>
    <!-- 为 ECharts 准备一个具备大小(宽高)的 Dom -->

<!--
JDBC 驱动名及数据库 URL
数据库的用户名与密码，根据自己的需要设置
useUnicode=true&characterEncoding=utf-8 防止中文乱码
```

```
-->
<sql:setDataSource var="snapshot" driver="com.mysql.jdbc.Driver"
    url="jdbc:mysql://localhost:3306/chubanshe?&characterEncoding=utf-8"
    user="root"  password="66688888"/>

<sql:query dataSource="${snapshot}" var="result">
SELECT count(bookname) as num1 FROM 'javabooks' WHERE data > '2018-01-30';
</sql:query>
<sql:query dataSource="${snapshot}" var="result1">
SELECT count(bookname) as num2 FROM 'pythonbooks' WHERE data > '2018-01-30';
</sql:query>
<c:forEach var="row" items="${result.rows}">

   <div id="main" style="width: 600px;height:400px;"></div>
   <script type="text/javascript">
       // 基于准备好的Dom，初始化Echarts实例
       var myChart = echarts.init(document.getElementById('main'));

       // 指定图表的配置项和数据
       var option = {
           title: {
               text: '××出版社最近Java和Python对比'
           },
           tooltip: {},
           xAxis: {
               data: ["Java","Python"]
           },
           yAxis: {},
           series: [{
               name: '此类书数量',
               type: 'bar',
               data: [<c:out value="${row.num1}"/>,
                  </c:forEach>
                  <c:forEach var="row" items="${result1.rows}">
                  <c:out value="${row.num2}"/>]
           }]
                  </c:forEach>
       };
               // 使用刚指定的配置项和数据显示图表
       myChart.setOption(option);
   </script>
</body>
```

文件执行效果如图7-32所示。由此可见，2023年出版的Python图书数量要多于Java图书。

第 7 章 图书市场数据分析系统

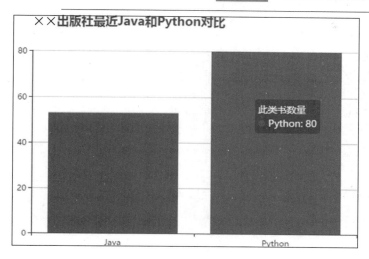

图 7-32 Java 图书和 Python 图书数量对比

7.5.4 大数据分析并可视化主分类图书数据

编写文件 BookDuibi.jsp，功能是提取数据库中的电子、经济、生活、摄影、计算机、科普和管理共计 7 个主分类图书的信息，然后大数据可视化展示这 7 类书的条形图数据。文件 BookDuibi.jsp 的主要实现代码如下。

```
<sql:query dataSource="${snapshot}" var="result">
select t1.num1,t2.num2,t3.num3,t4.num4,t5.num5,t6.num6,t7.num7 from (select
count(bookName) num1 from dianzibook) t1, (select count(bookName) num2 from
jingjibooks) t2, (select count(bookName) num3 from shenghuo) t3, (select
count(bookName) num4 from sheyingbook) t4, (select count(bookName) num5 from
jsjbooks) t5, (select count(bookName) num6 from kepuBook) t6, (select count(bookName)
num7 from guanlibook) t7;
</sql:query>


    <div id="container" style="height: 100%"></div>
    <script type="text/javascript"
src="http://echarts.baidu.com/gallery/vendors/echarts/echarts.min.js"></script>
    <script type="text/javascript" src="http://echarts.baidu.com/gallery/
vendors/echarts-gl/echarts-gl.min.js"></script>
    <script type="text/javascript" src="http://echarts.baidu.com/gallery/
vendors/echarts-stat/ecStat.min.js"></script>
    <script type="text/javascript" src="http://echarts.baidu.com/gallery/
vendors/echarts/extension/dataTool.min.js"></script>
```

```html
        <script type="text/javascript" src="http://echarts.baidu.com/gallery/vendors/echarts/map/js/china.js"></script>
        <script type="text/javascript" src="http://echarts.baidu.com/gallery/vendors/echarts/map/js/world.js"></script>
        <script type="text/javascript" src="http://echarts.baidu.com/gallery/vendors/echarts/extension/bmap.min.js"></script>
        <script type="text/javascript" src="http://echarts.baidu.com/gallery/vendors/simplex.js"></script>

        <c:forEach var="row" items="${result.rows}">
        <div id="main" style="width: 600px;height:400px;">
        <script type="text/javascript">
```
```javascript
var dom = document.getElementById("container");
var myChart = echarts.init(dom);
var app = {};
option = null;
app.title = '出版社图书 - 条形图';

option = {
    title: {
        text: '出版社主流图书对比',
        subtext: '2023年3月1日数据'
    },
    tooltip: {
        trigger: 'axis',
        axisPointer: {
            type: 'shadow'
        }
    },
    legend: {
        data: ['2023年3月1日数据']
    },
    grid: {
        left: '3%',
        right: '4%',
        bottom: '3%',
        containLabel: true
    },
    xAxis: {
        type: 'value',
        boundaryGap: [0, 0.01]
    },
    yAxis: {
        type: 'category',
        data: ['电子','经济','生活','摄影','计算机','科普','管理']
    },
    series: [
```

```
            {
                name: '2023年3月1日',
                type: 'bar',
                data: [<c:out value="${row.num1}"/>, <c:out value="${row.num2}"/>,
<c:out value="${row.num3}"/>, <c:out value="${row.num4}"/>, <c:out value=
"${row.num5}"/>, <c:out value="${row.num6}"/>, <c:out value="${row.num7}"/>]
            }
        ]
};
;
if (option && typeof option === "object") {
    myChart.setOption(option, true);
}
        </script>
            </c:forEach>
            </div>
        </body>
```

文件执行效果如图 7-33 所示。

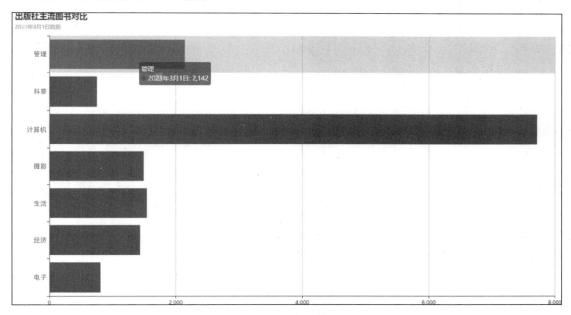

图 7-33　主分类图书数据对比

7.5.5　大数据分析并可视化计算机子类图书数据

编写文件 JSJBing.jsp，功能是提取数据库中 Java、Python、多媒体、考试、移动开发、办公、辅助设计、图形图像、通信、其他等类型图书的数据,然后大数据可视化展示这 10 类数据所占百分比的饼形图。文件 JSJBing.jsp 的主要实现代码如下。

```
<script src="echarts.common.min.js"></script>
<title>计算机图书统计饼形图</title>
</head>
  <body style="height: 100%; margin: 0">
    <div id="container" style="height: 100%"></div>
    <script type="text/javascript" src="echarts.min.js"></script>
    <script type="text/javascript" src="http://echarts.baidu.com/gallery/
vendors/echarts-gl/echarts-gl.min.js"></script>
<sql:setDataSource var="snapshot" driver="com.mysql.jdbc.Driver"
    url="jdbc:mysql://localhost:3306/chubanshe?&characterEncoding=utf-8"
    user="root"  password="66688888"/>

<sql:query dataSource="${snapshot}" var="result">
select t1.num1,t2.num2,t3.num3,t4.num4,t5.num5,t6.num6,t7.num7,t8.num8,t9.num9,
t10.num10 from (select count(bookName) num1 from javabooks) t1, (select count
(bookName) num2 from pythonbooks) t2, (select count(bookName) num3 from
jsjduomeitibooks) t3, (select count(bookName) num4 from jsjkaoshibooks) t4, (select
count(bookName) num5 from jsjmobilebooks) t5, (select count(bookName) num6 from
jsjofficebooks) t6, (select count(bookName) num7 from jsjshejibooks) t7, (select
count(bookName) num8 from jsjbooks) t8, (select count(bookName) num9 from
tuxingbooks) t9, (select count(bookName) num10 from jsjtongxinbooks) t10
</sql:query>

<sql:query dataSource="${snapshot}" var="result1">
SELECT (select count(bookName) total from jsjbooks)-(select count(bookName) total
from javabooks)-(select count(bookName) total from pythonbooks)-(select
count(bookName) total from jsjduomeitibooks)-(select count(bookName) total from
jsjkaoshibooks)-(select count(bookName) total from jsjmobilebooks)-(select
count(bookName) total from jsjofficebooks)-(select count(bookName) total from
jsjshejibooks)-(select count(bookName) total from tuxingbooks)-(select
count(bookName) total from jsjtongxinbooks) AS SumCount
</sql:query>


<c:forEach var="row" items="${result.rows}">

<script type="text/javascript">
var dom = document.getElementById("container");
var myChart = echarts.init(dom);
var app = {};
option = null;
option = {
   title : {
      text: '计算机图书统计饼形图',
```

```
            subtext: '计算机图书总计<c:out value="${row.num8}"/>本',
            x:'center'
        },
        tooltip : {
            trigger: 'item',
            formatter: "{a} <br/>{b} : {c} ({d}%)"
        },
        legend: {
            orient: 'vertical',
            left: 'left',
            data: ["Java","Python","多媒体","考试","移动开发","办公","辅助设计","图形图像","通信","其他"]
        },
        series : [
            {
                name: '分类所占比例',
                type: 'pie',
                radius : '55%',
                center: ['50%', '60%'],
                data:[
                    {value:<c:out value="${row.num1}"/>, name:'Java'},
                    {value:<c:out value="${row.num2}"/>, name:'Python'},
                    {value:<c:out value="${row.num3}"/>, name:'多媒体'},
                    {value:<c:out value="${row.num4}"/>, name:'考试'},
                    {value:<c:out value="${row.num5}"/>, name:'移动开发'},
                    {value:<c:out value="${row.num6}"/>, name:'办公'},
                    {value:<c:out value="${row.num7}"/>, name:'辅助设计'},
                    {value:<c:out value="${row.num9}"/>, name:'图形图像'},
                    {value:<c:out value="${row.num10}"/>, name:'通信'},
                    </c:forEach>
                    <c:forEach var="row" items="${result1.rows}">
                    {value:<c:out value="${row.SumCount}"/>, name:'其他'}
                    </c:forEach>
                ],
                itemStyle: {
                    emphasis: {
                        shadowBlur: 10,
                        shadowOffsetX: 0,
                        shadowColor: 'rgba(0, 0, 0, 0.5)'
                    }
                }
            }
        ]
};
;
if (option && typeof option === "object") {
    myChart.setOption(option, true);
```

```
}
    </script>

</body>
```

文件执行效果如图 7-34 所示。

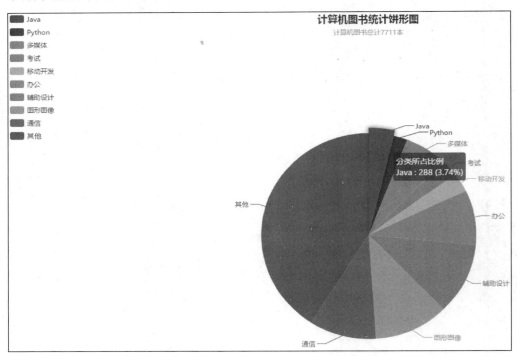

图 7-34 执行效果

第 8 章

基于深度学习的音乐推荐系统

音乐推荐系统是指根据用户的喜好推荐音乐,帮助用户在海量的音乐库中快速并精准地找到个人喜欢的音乐。本章将详细讲解使用 Java 语言开发音乐推荐系统的方法,展示深度学习技术在 Java 项目中的应用。本章项目由 TensorFlow +Spring+MyBatis+MySQL 实现。

8.1 背景介绍

信息技术的普及和发展改变了人们的生活方式，人们对娱乐的需求越来越强烈。各种音乐推荐系统应运而生。信息技术与互联网技术相结合的音乐网站，其便利性、数据存储安全性、共享性、数据容量等明显优于传统的磁带与CD。

扫码看视频

设计本系统的目的是改善那些原有网站的缺陷，为用户提供一个性能更好、使用更便利的在线音乐系统，并且在协调用户的偏好方面也可以做得更好。此外，系统能提供一个清晰简明的界面，合理安排音乐分类信息，根据用户的某些特点，提供更能产生共鸣的音乐。

8.2 系统分析

本系统采用友好的界面，为用户推荐适合自己的音乐。具体推荐方法是通过深度学习技术，根据个人喜好进行算法分析，得到每一位用户喜好的音乐。

扫码看视频

8.2.1 系统功能分析

随着移动网络和数字多媒体技术的发展，数字音乐产业的共享与广泛传播得到了进一步发展。对用户而言，在海量的音乐库中寻找喜欢的音乐需要花费大量的时间和精力。音乐推荐系统的目的是将用户从这项烦琐的工作中解脱出来，有效地提升用户体验，并为音乐平台创造经济收益。

8.2.2 系统需求分析

本音乐推荐系统的目标是以个性化音乐推荐模型为基础，使用 B/S 架构实现个性化音乐推荐。在本系统中，用户可以浏览音乐，可以收藏音乐，可以为所喜爱的音乐点赞，同时还可以进行用户登录和注册。管理员除了可以实现普通用户所具有的功能外，还可以进行音乐、评论、用户的管理。系统通过隐式收集用户操作记录，向用户推荐个性化的音乐。与此同时，该系统还具有排行榜、热门推荐等普通音乐网站所具有的功能，其中排行榜分为周榜和月榜，热门推荐是将当天用户点击量最高的 50 首歌曲推荐给用户。

8.2.3 系统模块分析

- 用户管理模块：包括用户注册、登录、收藏、评论、点赞、浏览历史记录、搜索音乐、播放控制音乐、音乐下载等功能。
- 管理员管理模块：包括用户查找、用户删除、音乐上传、评论查询、评论删除、歌曲查询、歌曲删除等功能。
- 排行榜模块：包括周榜排行和月榜排行，分别表示当周和当月播放量最高的歌曲推荐。
- 热门推荐模块：对播放量、用户评论量等数据进行综合分析，得出一些热门歌曲推荐给用户。
- 个性化推荐模块：通过协同过滤推荐算法收集用户操作数据，从而进行音乐推荐；通过分析用户播放歌曲的歌词数据，结合深度学习领域相关算法对用户进行歌曲推荐。

本音乐推荐系统的模块结构如图 8-1 所示。

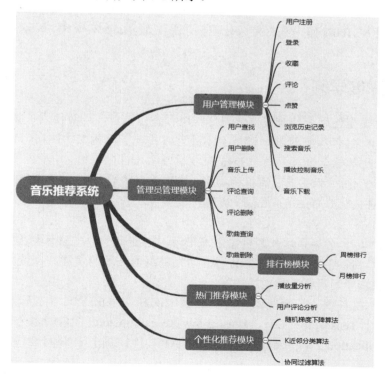

图 8-1 系统模块结构图

8.3 系统架构分析

本系统采用 MVC 模式架构，并使用了深度学习技术和核心算法知识，最终实现个性化音乐推荐系统。

扫码看视频

8.3.1 MVC 架构

1) 前端架构

本系统前端基于 Bootstrap 框架实现，采用了 JSP、JavaScript、CSS 技术进行开发。

2) 后端架构

本系统后端采用 SSM(Spring+SpringMVC+MyBatis)框架进行开发，该框架为当前企业中较为流行的一种框架。SSM 框架集由 Spring、MyBatis 两个开源框架整合而成(SpringMVC 是 Spring 中的部分内容)，常作为数据源较简单的 Web 项目的框架。

3) 数据持久化架构

本系统使用 MyBatis 作为数据持久化框架，通过 MyBatis 实现和 MySQL 数据库数据的映射。

8.3.2 深度学习

本系统推荐模型使用了传统机器学习算法进行音乐推荐，同时使用了类似于 Word2vec 的词袋模型和词向量模型来对歌词进行文本处理，构建了异构文本网络，来标识用户的歌曲偏好，然后在此基础上引入了一个 Java 深度学习库 deepLearning4j 对音乐特征进行提取，并对音乐进行标签化(分为古典、流行等类别，可以进行混合推荐)。本系统使用著名的深度学习框架 TensorFlow 和 Deeplearning4j 实现。

1) TensorFlow

TensorFlow 是一个端到端开源机器学习平台。它拥有一个全面而灵活的生态系统，其中包含各种工具、库和社区资源，可助力推动先进机器技术的发展，并使开发者能够轻松构建和部署由机器学习提供支持的应用。

TensorFlow 由谷歌人工智能团队谷歌大脑(Google Brain)负责开发和维护，拥有包括 TensorFlow Hub、TensorFlow Lite、TensorFlow Research Cloud 在内的多个项目以及各类应用程序接口(Application Programming Interface, API)。自 2015 年 11 月 9 日起，TensorFlow 依据 Apache 2.0 协议开放源代码。

TensorFlow 的 Java 版本支持 Windows、Mac OS、Linux、Android 等主流操作系统。本

项目需要先引用库 tensorflow-1.5.0.jar，之后才能使用 TensorFlow 的功能。

2) Deeplearning4j

Deeplearning4j(简称 DL4J)是为 Java 和 Scala 编写的首个商业级开源分布式深度学习库。DL4J 与 Hadoop 和 Spark 集成，为商业环境(而非研究工具目的)设计。Deeplearning4j 技术先进，以即插即用为目标，通过更多预设的使用，避免太多配置，让非研究人员也能够进行快速的原型制作。DL4J 同时可以规模化定制。DL4J 遵循 Apache 2.0 许可协议，一切以其为基础的衍生作品均属于衍生作品的作者。

8.4 数据库设计

数据库设计是系统设计开发的第一步，项目是否成功主要看数据设计是否合理，数据库设计的好坏直接影响程序编码的复杂程度。本系统使用 MyBatis 和 MySQL 数据库实现数据的持久化。

扫码看视频

8.4.1 数据库架构设计

本项目用到了 11 个数据库表和 3 个视图，具体结构如图 8-2 所示。

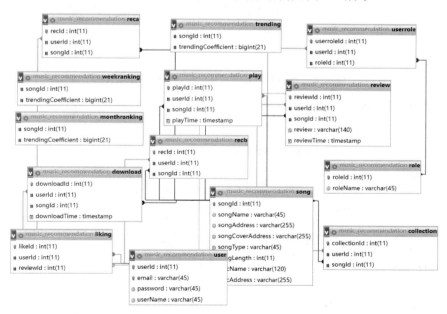

图 8-2 数据库架构设计

8.4.2 数据库结构设计

本项目用到了 11 个数据库表和 3 个视图。创建名为 music_recommendation 的数据库，然后设计各个数据库表的结构，具体说明如下。

(1) 歌曲收藏表(collection)用于保存用户的收藏信息，具体设计结构如图 8-3 所示。

名字	类型	排序规则	属性	空	默认	注释	额外
collectionId	int(11)			否	无		AUTO_INCREMENT
userId	int(11)			否	无		
songId	int(11)			否	无		

图 8-3 歌曲收藏表(collection)的设计结构

(2) 歌曲下载记录表(download)用于保存用户下载歌曲的信息，具体设计结构如图 8-4 所示。

名字	类型	排序规则	属性	空	默认	注释	额外
downloadId	int(11)			否	无		AUTO_INCREMENT
userId	int(11)			否	无		
songId	int(11)			否	无		
downloadTime	timestamp			否	CURRENT_TIMESTAMP		

图 8-4 歌曲下载记录表(download)的设计结构

(3) 歌曲推荐表 a(reca)用于保存为用户推荐的 a 版歌曲信息，具体设计结构如图 8-5 所示。

名字	类型	排序规则	属性	空	默认	注释	额外
recId	int(11)			否	无		AUTO_INCREMENT
userId	int(11)			否	无		
songId	int(11)			否	无		

图 8-5 歌曲推荐表 a(reca)的设计结构

(4) 歌曲推荐表 b(reca)用于保存为用户推荐的 b 版歌曲信息，具体设计结构如图 8-6 所示。

(5) 歌曲评论表(review)用于保存用户为某歌曲发布的评论信息，具体设计结构如图 8-7 所示。

名字	类型	排序规则	属性	空	默认	注释	额外
recId	int(11)			否	无		AUTO_INCREMENT
userId	int(11)			否	无		
songId	int(11)			否	无		

图 8-6 歌曲推荐表 b(reca)的设计结构

名字	类型	排序规则	属性	空	默认	注释	额外
reviewId	int(11)			否	无		AUTO_INCREMENT
userId	int(11)			否	无		
songId	int(11)			否	无		
review	varchar(140)	utf8_general_ci		否	无		
reviewTime	timestamp			否	CURRENT_TIMESTAMP		

图 8-7 歌曲评论表(review)的设计结构

(6) 用户信息表(user)用于保存系统内的用户信息，包括普通会员和管理员，具体设计结构如图 8-8 所示。

名字	类型	排序规则	属性	空	默认	注释	额外
userId	int(11)			否	无		AUTO_INCREMENT
email	varchar(45)	utf8_general_ci		否	无		
password	varchar(45)	utf8_general_ci		否	无		
userName	varchar(45)	utf8_general_ci		是	毛毛		

图 8-8 用户信息表(user)的设计结构

(7) 用户权限信息表(role)用于保存系统内用户的权限信息，如果 roleId 值是 1，表示这是管理员。具体设计结构如图 8-9 所示。

名字	类型	排序规则	属性	空	默认	注释	额外
roleId	int(11)			否	无		AUTO_INCREMENT
roleName	varchar(45)	utf8_general_ci		否	无		

图 8-9 用户权限信息表(role)的设计结构

(8) 用户角色信息表(userrole)用于保存系统内的用户角色信息，设置每一位用户的权限。具体设计结构如图 8-10 所示。

(9) 音乐详情信息表(song)用于保存系统内的音乐信息，具体设计结构如图 8-11 所示。

名字	类型	排序规则	属性	空	默认	注释	额外
userroleId	int(11)			否	无		AUTO_INCREMENT
userId	int(11)			否	无		
roleId	int(11)			否	无		

图 8-10　用户角色信息表(userrole)的设计结构

名字	类型	排序规则	属性	空	默认	注释	额外
songId	int(11)			否	无		AUTO_INCREMENT
songName	varchar(45)	utf8_general_ci		否	无		
songAddress	varchar(255)	utf8_general_ci		否	无		
songCoverAddress	varchar(255)	utf8_general_ci		是	NULL		
songType	varchar(45)	utf8_general_ci		是	NULL		
songLength	int(11)			否	0		
lyricName	varchar(120)	utf8_general_ci		是	NULL		
lyricAddress	varchar(255)	utf8_general_ci		是	NULL		

图 8-11　音乐详情信息表(song)的设计结构

(10) 音乐播放记录信息表(play)用于保存系统内音乐的播放信息，具体设计结构如图 8-12 所示。

名字	类型	排序规则	属性	空	默认	注释	额外
playId	int(11)			否	无		AUTO_INCREMENT
userId	int(11)			否	无		
songId	int(11)			否	无		
playTime	timestamp			否	CURRENT_TIMESTAMP		

图 8-12　音乐播放记录信息表(play)的设计结构

(11) 用户点赞信息表(liking)用于保存系统内用户对指定音乐的点赞信息，具体设计结构如图 8-13 所示。

名字	类型	排序规则	属性	空	默认	注释	额外
likeId	int(11)			否	无		AUTO_INCREMENT
userId	int(11)			否	无		
reviewId	int(11)			否	无		

图 8-13　用户点赞信息表(liking)的设计结构

(12) 系统会根据统计用户的播放记录，以周为单位进行排行，并将排行信息保存到视图 weekranking 中。具体设计结构如图 8-14 所示。

名字	类型	排序规则	属性	空	默认
songId	int(11)			否	无
trendingCoefficient	bigint(21)			否	0

图 8-14 周榜排行信息视图(weekranking)的设计结构

（13）系统会根据统计用户的播放记录，以月为单位进行排行，将排行信息保存到视图 monthranking 中。具体设计结构如图 8-15 所示。

名字	类型	排序规则	属性	空	默认
songId	int(11)			否	无
trendingCoefficient	bigint(21)			否	0

图 8-15 月榜排行信息视图(monthranking)的设计结构

（14）系统会将一些热门歌曲推荐给用户，这些推荐歌曲是根据歌曲的播放量、用户的评论量等数据进行综合分析得出的，将热门歌曲信息保存到视图 trending 中。具体设计结构如图 8-16 所示。

名字	类型	排序规则	属性	空	默认
songId	int(11)			否	无
trendingCoefficient	bigint(21)			否	0

图 8-16 热门歌曲信息视图(trending)的设计结构

8.5 用户管理模块

本项目的用户管理模块包括用户注册、登录、收藏、评论、点赞、浏览历史记录、搜索音乐、播放控制音乐、下载等功能。本节将详细讲解用户管理模块的实现过程。

扫码看视频

8.5.1 用户注册

编写文件 RegisterController.java，功能是获取用户的邮箱账号信息、验证码信息和密码信息，如果合法则注册成功。文件 RegisterController.java 的具体实现代码如下。

```
@Controller
public class RegisterController {
    @Autowired
    private UserService userService;
```

```java
@Autowired
private PersonalRecService personalRecService;

@PostMapping(value = "getValidateCode.do",produces = "text/html;charset=UTF-8")
@ResponseBody
public String getValidateCode(HttpServletRequest request,String email) {
    //邮箱是否已经存在
    boolean isExisted=userService.isEmailExisted(email);
    if(isExisted) {
        return ReturnMsg.msg(HttpServletResponse.SC_BAD_REQUEST, "该用户已存在");
    }
    //防止疯狂发送，这里简单地限制一下，1分钟只发送一次
    if(userService.tooQuickly(request,1)) {
        return ReturnMsg.msg(HttpServletResponse.SC_BAD_REQUEST, "发送频率太快");
    }
    //subject
    String subject="验证码验证";
            //validation code
    String code=(int)(Math.random()*10000)+"";
    String content="非常高兴您能加入我们，您本次的验证码为："+code+"\n\n"
                +"再次感谢您的加入";
    boolean isSuccessful=MailService.sendEmail(email,subject,content);
    if(isSuccessful) {
        request.getSession().setAttribute("code", code);
        return ReturnMsg.msg(HttpServletResponse.SC_OK, "发送成功");
    }else {
        return ReturnMsg.msg(HttpServletResponse.SC_BAD_REQUEST, "发送失败");
    }
}


@PostMapping(value = "register.do",produces = "text/html;charset=UTF-8")
@ResponseBody
public String register(HttpServletRequest request,User u) {
    //验证码是否正确
    String code=(String) request.getSession().getAttribute("code");
    if(code==null || !code.equals(u.getValidateCode())) {
        return ReturnMsg.msg(HttpServletResponse.SC_BAD_REQUEST, "验证码错误");
    }
    boolean isInserted=userService.insert(u);
    if(isInserted) {
        request.getSession().setAttribute("user", u);
        /**
         * 用户注册成功时，初始化个性推荐列表
         */
        personalRecService.initializePersonalRecList(request);
```

```java
                return ReturnMsg.msg(HttpServletResponse.SC_OK,
                                JSONObject.toJSON(u).toString());
        }else {
            return ReturnMsg.msg(HttpServletResponse.SC_BAD_REQUEST, "注册失败");
        }
    }
}
```

用户注册时会向用户的邮箱中发送验证码，编写文件 MailService.java 实现发送验证码功能，具体代码如下。

```java
public class MailService {
    @Autowired
    private JavaMailSender sender;
    //用户名
    private final static String from="@163.com";
    //密码
    private final static String psw=" ";
    //邮箱服务器
    private final static String host="smtp.163.com";
    /**
     * 发送纯文本的简单邮件
     * @param to
     * @param subject
     * @param content
     */
    public void sendSimpleMail(String to, String subject, String content){
        SimpleMailMessage message = new SimpleMailMessage();
        message.setFrom(from);
        message.setTo(to);
        message.setSubject(subject);
        message.setText(content);

        try {
            sender.send(message);
            System.err.println("简单邮件已经发送。");
        } catch (Exception e) {
            System.err.println("发送简单邮件时发生异常！"+e);
        }
    }

    /**
     * 使用加密的方式，利用 465 端口传输邮件，开启 ssl
     * @param to        为收件人邮箱
     * @param message   发送的消息
     */
    public static boolean sendEmail(String to,String subject, String message) {
```

```java
    try {
        Security.addProvider(new com.sun.net.ssl.internal.ssl.Provider());
        final String SSL_FACTORY = "javax.net.ssl.SSLSocketFactory";
        //设置邮件会话参数
        Properties props = new Properties();
        //邮箱的发送服务器地址
        props.setProperty("mail.smtp.host", host);
        props.setProperty("mail.smtp.socketFactory.class", SSL_FACTORY);
        props.setProperty("mail.smtp.socketFactory.fallback", "false");
        //邮箱发送服务器端口,这里设置为465端口
        props.setProperty("mail.smtp.port", "465");
        props.setProperty("mail.smtp.socketFactory.port", "465");
        props.put("mail.smtp.auth", "true");
        final String username = from;
        final String password = psw;
        //获取到邮箱会话,利用匿名内部类的方式,将发送者邮箱用户名和密码授权给jvm
        Session session = Session.getDefaultInstance(props, new Authenticator() {
            protected PasswordAuthentication getPasswordAuthentication() {
                return new PasswordAuthentication(username, password);
            }
        });
        //通过会话得到一个邮件,用于发送
        Message msg = new MimeMessage(session);
        //设置发件人
        msg.setFrom(new InternetAddress(from));
        //设置收件人,to为收件人,cc为抄送,bcc为密送
        msg.setRecipients(Message.RecipientType.TO, InternetAddress.parse(to, false));
        msg.setRecipients(Message.RecipientType.CC, InternetAddress.parse(to, false));
        msg.setRecipients(Message.RecipientType.BCC, InternetAddress.parse(to, false));
        msg.setSubject(subject);
        //设置邮件消息
        msg.setContent(message,"text/html;charset=utf-8");
        //msg.setText(message);
        // msg.setContent(text,"text/html;charset=UTF-8");
        //设置发送的日期
        msg.setSentDate(new Date());
        //调用Transport的send方法发送邮件
        Transport.send(msg);
        return true;
    } catch (Exception e) {
        e.printStackTrace();
        return false;
    }
}
```

用户注册成功后,系统会将注册信息添加到数据库中。为了提高系统的安全性,将密

码信息使用 MD5 技术加密。文件 MD5Util.java 用于实现密码加密功能，具体实现代码如下。

```java
public class MD5Util {

    /**
     * MD5 生成 32 位的 md5 码
     * @param inStr
     * 需要加密的字符串
     * @return
     */
    public static String string2MD5(String inStr){
        MessageDigest md5 = null;
        try{
            md5 = MessageDigest.getInstance("MD5");
        }catch (Exception e){
            System.out.println(e.toString());
            e.printStackTrace();
            return "";
        }
        char[] charArray = inStr.toCharArray();
        byte[] byteArray = new byte[charArray.length];

        for (int i = 0; i < charArray.length; i++)
            byteArray[i] = (byte) charArray[i];
        byte[] md5Bytes = md5.digest(byteArray);
        StringBuffer hexValue = new StringBuffer();
        for (int i = 0; i < md5Bytes.length; i++){
            int val = ((int) md5Bytes[i]) & 0xff;
            if (val < 16)
                hexValue.append("0");
            hexValue.append(Integer.toHexString(val));
        }
        return hexValue.toString();
    }
}
```

8.5.2 用户登录

编写文件 LoginController.java，功能是获取用户的登录信息，如果合法则登录成功。文件 LoginController.java 的具体实现代码如下。

```java
public class LoginController {
    @Autowired
    private UserService userService;
    @PostMapping(value = "login.do",produces = "text/html;charset=UTF-8")
    @ResponseBody
```

```java
public String login(HttpServletRequest request, User u) {
    boolean isUserExisted=userService.findLogin(u);
    if(!isUserExisted) {
        return ReturnMsg.msg(HttpServletResponse.SC_BAD_REQUEST, "账号或密码错误");
    }else {
        request.getSession().setAttribute("user", u);
        request.getSession().setAttribute("isHasPrivilege",
                userService.isHasPrivilege(request));
        return ReturnMsg.msg(HttpServletResponse.SC_OK,
                JSONObject.toJSON(u).toString());
    }
}
```

8.5.3 收藏歌曲

当用户登录系统后，可以随时收藏喜欢的歌曲。编写文件 CollectionController.java，实现收藏业务逻辑，具体实现代码如下。

```java
public class CollectionController {
    @Autowired
    private CollectionService collectionService;

    //只接受 post 方式的请求
    @PostMapping(value = "collectSong.do",produces = "text/html;charset=UTF-8")
    /**
     * @ResponseBody
     * 注解的作用是将 controller 的方法返回的对象通过适当的转换器转换为指定的格式之后，
     * 写入 response 对象的 body 区，通常用来返回 JSON 数据或者是 XML
     */
    @ResponseBody
    public String collectSong(HttpServletRequest request,int songId) {
        boolean isCollected=collectionService.collectionChange(request,songId);
        return ReturnMsg.msg(HttpServletResponse.SC_OK, isCollected+"");
    }
}
```

编写文件 CollectionServiceImpl.java，实现具体的收藏功能，具体实现代码如下。

```java
@Service("collectionService")
public class CollectionServiceImpl implements CollectionService {
    @Autowired
    private CollectionDao collectionDao;
    @Autowired
    private UserDao userDao;
```

```java
public boolean collectionChange(HttpServletRequest request, int songId) {
    boolean isCurCollected=true;
    User user=userDao.selectByUser(Request.getUserFromHttpServletRequest(request));
    Collection collection=collectionDao.selectByCollection(new
             Collection(user.getUserId(),songId));
    if(collection==null) {
        //该歌曲没有被收藏
        isCurCollected=false;
        //添加收藏
        collectionDao.insert(new Collection(user.getUserId(),songId));
    }else {
        //若已经被收藏，则取消收藏
        collectionDao.deleteById(collection.getCollectionId());
    }
    //返回改变后的收藏状态
    return !isCurCollected;
}
public List<Collection> getAllRecords() {
    return collectionDao.selectAll();
}
}
}
```

8.5.4 用户评论和点赞

编写文件 ReviewController.java，实现用户评论功能。可以在表单中发布评论，发布成功后将评论添加到数据库中。文件 ReviewController.java 的具体实现代码如下。

```java
public class ReviewController {
    @Autowired
    private SongService songService;
    @Autowired
    private ReviewService reviewService;
    @Autowired
    private UserService userService;

    @RequestMapping(value = "reviewFrameLoad.do", method = { RequestMethod.GET })
    public ModelAndView reviewFrameLoad(HttpServletRequest request, int songId) {
        //获取当前选中的歌曲
        Song song = songService.getSongByIdWithCollectionFlag(request, songId);
        //获取选中歌曲的精彩评论
        List<Review> hotReviewList =
             reviewService.getHotReviewBySongIdWithLikeFlag(request,songId);
        //获取选中歌曲的最新评论(目前评论数据很少，先不做分页)
        List<Review> newReviewList =
             reviewService.getNewReviewBySongIdWithLikeFlag(request,songId);
```

```java
        ModelAndView modelAndView = new ModelAndView();
        modelAndView.setViewName("reviewFrame");

        modelAndView.addObject("song", song);
        modelAndView.addObject("hotReviewList", hotReviewList);
        modelAndView.addObject("newReviewList", newReviewList);

        return modelAndView;
    }

    @PostMapping(value = "review.do")
    @ResponseBody
    public String review(HttpServletRequest request,int songId,String review) {
        boolean isAdded=reviewService.addReview(request,songId,review);
        if(isAdded) {
            return ReturnMsg.msg(HttpServletResponse.SC_OK,"评论成功");
        }
        return ReturnMsg.msg(HttpServletResponse.SC_BAD_REQUEST, "评论出错");
    }

    @GetMapping(value = "reviewLike.do",produces = "text/html;charset=UTF-8")
    @ResponseBody
    public String reviewLike(HttpServletRequest request,int reviewId) {
        boolean isLiked=reviewService.reviewLikeChange(request,reviewId);
        return ReturnMsg.msg(HttpServletResponse.SC_OK, isLiked+"");
    }

    @RequestMapping(value = "newReviewFrameLoad.do", method = { RequestMethod.GET })
    public ModelAndView newReviewFrameLoad(HttpServletRequest request, int songId) {
        //获取选中歌曲的最新评论(目前评论数据很少,先不做分页)
        List<Review> newReviewList =
                reviewService.getNewReviewBySongIdWithLikeFlag(request,songId);

        ModelAndView modelAndView = new ModelAndView();
        modelAndView.setViewName("newReviewFrame");
        modelAndView.addObject("newReviewList", newReviewList);

        return modelAndView;
    }

    @RequestMapping(value = "deleteReview.do", method = { RequestMethod.POST })
    public void deleteReview(HttpServletRequest request, int reviewIds[]) {
        if(userService.isHasPrivilege(request)) {
            reviewService.batchDeleteById(reviewIds);
        }
    }
}
```

编写文件 ReviewServiceImpl.java，实现用户点赞功能。首先获取用户的点赞列表，然后将点赞的信息添加到列表中。文件 ReviewServiceImpl.java 的具体实现代码如下。

```java
@Service("reviewServiceImpl")
public class ReviewServiceImpl implements ReviewService {
    @Autowired
    private UserDao userDao;
    @Autowired
    private ReviewDao reviewDao;

    public boolean addReview(HttpServletRequest request, int songId, String content) {
        boolean isInsertSuccessful = false;
        User user = userDao.selectByUser(Request.getUserFromHttpServletRequest
                (request));
        if (user == null) {
            return isInsertSuccessful;
        }
        Review review = new Review(user.getUserId(), songId, content);
        int affectedRows = reviewDao.insert(review);
        if (affectedRows > 0) {
            isInsertSuccessful = true;
        }
        return isInsertSuccessful;
    }

    public List<Review> getHotReviewBySongIdWithLikeFlag(HttpServletRequest
        request, int songId) {
        User user = userDao.selectByUser(Request.getUserFromHttpServletRequest
                (request));
        //获取用户的点赞列表
        List<Like> likeList=null;
        if(user!=null) {
            likeList= reviewDao.selectLikeByUserId(user.getUserId());
        }
        //获取歌曲的精彩评论列表
        List<Review> hotReviewList= reviewDao.selectHotReviewWithLikeNumber(songId);

        //在结果列表中给已经被该用户点赞的评论加上标记
        if(hotReviewList!=null && likeList!=null) {
            for(Like like:likeList) {
                for(Review review:hotReviewList) {
                    if(like.getReviewId()==review.getReviewId()) {
                        review.setWhetherLiked(true);
                    }
                }
            }
        }
```

```java
        return hotReviewList;
    }

    public boolean reviewLikeChange(HttpServletRequest request, int reviewId) {
        boolean isLiked=true;
        User user=userDao.selectByUser(Request.getUserFromHttpServletRequest
                (request));
        //获取当前评论的点赞状态
        Like like=reviewDao.selectByLike(new Like(user.getUserId(),reviewId));
        if(like==null) {
            //该评论还没有被该用户点赞
            isLiked=false;
            //进行点赞
            reviewDao.insertLikeRecord(new Like(user.getUserId(),reviewId));
        }else {
            //如果已经点赞了，则取消点赞
            reviewDao.deleteLikeRecordById(like.getLikeId());
        }
        //返回该评论改变后的点赞状态
        return !isLiked;
    }
    public List<Review> getNewReviewBySongIdWithLikeFlag(HttpServletRequest
                request, int songId) {
        User user = userDao.selectByUser(Request.getUserFromHttpServletRequest
                (request));
        //获取用户的点赞列表
        List<Like> likeList=null;
        if(user!=null) {
            likeList= reviewDao.selectLikeByUserId(user.getUserId());
        }
        //获取歌曲的最新评论列表
        List<Review> newReviewList= reviewDao.selectNewReviewWithLikeNumber
                (songId);
        //在结果列表中给已经被该用户点赞的评论加上标记
        if(newReviewList!=null && likeList!=null) {
            for(Like like:likeList) {
                for(Review review:newReviewList) {
                    if(like.getReviewId()==review.getReviewId()) {
                        review.setWhetherLiked(true);
                    }
                }
            }
        }
        return newReviewList;
    }
    public void batchDeleteById(int[] reviewIds) {
        if(reviewIds==null) {
```

```
            return;
        }
        reviewDao.deleteByIds(reviewIds);

    }
}
```

8.5.5 音乐播放记录

编写文件 RecordPlayController.java，记录用户的音乐播放信息，并将播放记录添加到数据库中。具体实现代码如下。

```
public class RecordPlayController {
    @Autowired
    private RecordPlayService recordPlayService;
    @GetMapping(value = "recordPlay.do")
    public void recordPlay(HttpServletRequest request,int songId) {
        recordPlayService.recordPlay(request,songId);

    }
}
```

8.5.6 音乐下载

编写文件 DownloadController.java，实现音乐下载功能，具体实现代码如下。

```
public class DownloadController {
    @Autowired
    private RecordDownloadService recordDownloadService;

    @RequestMapping(value = "download.do", method = { RequestMethod.GET})
    public void download(HttpServletRequest request,HttpServletResponse
                response,String songAddress,int songId) throws IOException {
        //对于登录用户，记录其下载信息
        recordDownloadService.recordDownload(request, songId);
        String songAddr = new String(songAddress.getBytes("utf-8"),
                "utf-8");
        response.setContentType("audio/mp3");
        response.setHeader("Content-Disposition", "attachment;filename="+
URLEncoder.encode(System.currentTimeMillis()+"如果不想返回名称的话.mp3", "utf-8"));
        BufferedOutputStream out = new BufferedOutputStream
                            (response.getOutputStream());
        InputStream bis=null;
        if(songAddr.contains("http")) {
            //在另外服务器的文件
```

```
            URL url = new URL(songAddr);
         URLConnection uc = url.openConnection();
            bis=new BufferedInputStream(uc.getInputStream());
        }else {
            //在服务器内部的文件
            songAddr=request.getServletContext().getRealPath(songAddr);
            
            bis = new BufferedInputStream(new FileInputStream(new File(songAddr)));
        }
        int len = 0;
        while((len = bis.read()) != -1){
         out.write(len);
         out.flush();
    }
        out.close();
        bis.close();
        
    }
}
```

8.6 管理员管理模块

系统管理员登录系统后,可以管理系统内的数据信息,包括用户查找、用户删除、音乐上传、评论查询、评论删除、歌曲查询、歌曲删除等功能。本节将详细讲解管理员管理模块的实现过程。

扫码看视频

8.6.1 信息搜索

编写文件 SearchController.java,实现信息搜索功能,包括音乐搜索、用户搜索和评论搜索;并且可以进一步对这些信息进行管理,例如删除操作。文件 SearchController.java 的具体实现代码如下。

```
public class SearchController {
   @Autowired
   private SearchService searchService;
   @Autowired
   private UserService userService;
   /**
    * @param mode
    * mode =0 :音乐搜索 ;
    * mode =1 :用户搜索 ;
    * mode =2 :评论搜索 ;
    * mode =null :
```

```java
 * @return
 */
@RequestMapping(value = "searchFrameLoad.do",method = { RequestMethod.GET })
public ModelAndView searchFrameLoad(HttpServletRequest request,String 
                                    keyword,String mode) {
    ModelAndView modelAndView=new ModelAndView();
    //管理员搜索
    if(mode!=null && userService.isHasPrivilege(request)) {
        int modeInt=Integer.parseInt(mode);
        if(modeInt==0) {
            //歌曲搜索
            modelAndView.setViewName("songManageSearchFrame");
            List<Song> songManageSearchList= searchService.getSearchSong
                    (keyword);
            modelAndView.addObject("songManageSearchList",songManageSearchList);
            if(songManageSearchList.size()==0) {
                modelAndView.addObject("oneDayOneWord","下落不明");
            }else {
                modelAndView.addObject("oneDayOneWord",
                    OneDayOneWord.getOneDayOneWord(Static.SEARCH_WORD_ARRAY));
            }
            
        }else if(modeInt==1) {
            //用户搜索
            modelAndView.setViewName("userManageSearchFrame");
            List<User> userManageSearchList=searchService.getSearchUser
                    (request,keyword);
            modelAndView.addObject("userManageSearchList", userManageSearchList);
            if(userManageSearchList.size()==0) {
                modelAndView.addObject("oneDayOneWord","下落不明");
            }else {
                modelAndView.addObject("oneDayOneWord",
                    OneDayOneWord.getOneDayOneWord(Static.SEARCH_WORD_ARRAY));
            }
            
        }else {
            //评论搜索
            modelAndView.setViewName("reviewManageSearchFrame");
            List<Review> reviewManageSearchList=
                        searchService.getSearchReview(keyword);
            modelAndView.addObject("reviewManageSearchList",
                    reviewManageSearchList);
            if(reviewManageSearchList.size()==0) {
                modelAndView.addObject("oneDayOneWord","下落不明");
            }else {
                modelAndView.addObject("oneDayOneWord",
                    OneDayOneWord.getOneDayOneWord(Static.SEARCH_WORD_ARRAY));
```

```
                    }
                }
            }else {
                modelAndView.setViewName("searchFrame");
                List<Song> searchSongList=
                    searchService.getSearchSongWithCollectionFlag(request,keyword);
                modelAndView.addObject("searchSongList",searchSongList);
                if(searchSongList.size()==0) {
                    modelAndView.addObject("oneDayOneWord","下落不明");
                }else {
                    modelAndView.addObject("oneDayOneWord",
                        OneDayOneWord.getOneDayOneWord(Static.SEARCH_WORD_ARRAY));
                }

            }

            return modelAndView;
        }
    }
```

编写文件 SearchServiceImpl.java，处理搜索结果，主要实现代码如下。

```
@Service("searchService")
public class SearchServiceImpl implements SearchService{
    @Autowired
    private SearchDao searchDao;
    @Autowired
    private UserDao userDao;
    @Autowired
    private TrendingRecDao trendingRecDao;
    public List<Song> getSearchSongWithCollectionFlag(HttpServletRequest request,
            String keyword) {
        List<Song> searchSongList=new ArrayList<Song>();
        List<Collection> collectionList=new ArrayList<Collection>();
        User user=userDao.selectByUser(Request.
                getUserFromHttpServletRequest(request));
        collectionList=trendingRecDao.getCollection(user);
        searchSongList=searchDao.selectSongLikeKeyword(keyword);

            //在搜索结果列表中给已经被该用户收藏的歌曲加上标记
            if(collectionList!=null && searchSongList!=null) {
                for(Collection c:collectionList) {
                    for(Song t:searchSongList) {
                        if(c.getSongId()==t.getSongId()) {
                            t.setWhetherCollected(true);
                        }
                    }
```

```
            }
        }
        return searchSongList;
    }
    public List<Review> getSearchReview(String keyword) {
        List<Review> searchReviewList=new ArrayList<Review>();
        searchReviewList=searchDao.selectReviewLikeKeyword(keyword);
        return searchReviewList;
    }
    public List<User> getSearchUser(HttpServletRequest request,String keyword) {
        User user=userDao.selectByUser
                   (Request.getUserFromHttpServletRequest(request));
        List<User> searchUserList=new ArrayList<User>();
        searchUserList=searchDao.selectUserLikeKeyword
                       (keyword,user.getUserId());
        return searchUserList;
    }
    public List<Song> getSearchSong(String keyword) {
        List<Song> searchSongList=new ArrayList<Song>();
        searchSongList=searchDao.selectSongLikeKeyword(keyword);
        return searchSongList;
    }
}
```

8.6.2 用户管理

编写文件 UserController.java，实现用户管理功能，管理员可以删除系统内的某个用户。文件 UserController.java 的具体实现代码如下。

```
public class UserController {
    @Autowired
    private UserService userService;

    @RequestMapping(value = "deleteUser.do", method = { RequestMethod.POST })
    public void deleteUser(HttpServletRequest request, int userIds[]) {
        if(userService.isHasPrivilege(request)) {
            userService.batchDeleteById(userIds);
        }
    }
}
```

8.6.3 音乐管理

编写文件 SongController.java，实现音乐管理功能，管理员可以向系统中添加新的音乐信息，也可以删除已经存在的音乐信息。文件 SongController.java 的具体实现代码如下。

```java
@Controller
public class SongController {
    @Autowired
    private UserService userService;
    @Autowired
    private SongService songService;

    @RequestMapping(value = "deleteSong.do", method = { RequestMethod.POST })
    public void deleteSong(HttpServletRequest request, int songIds[]) {
        if(userService.isHasPrivilege(request)) {
            songService.batchDeleteById(request,songIds);
        }

    }
    /**
     * 添加歌词歌曲
     */
    @PostMapping(value = "addSong.do",produces = "text/html;charset=UTF-8")
    @ResponseBody
    public String addSong(HttpServletRequest request, MultipartFile song,
                    MultipartFile lyric) {
        if(userService.isHasPrivilege(request) &&
            songService.checkFormat(song,lyric)) {
            boolean isSuccessful=songService.addSong(request,song,lyric);
            if(isSuccessful) {
                return ReturnMsg.msg(HttpServletResponse.SC_OK, "上传成功");
            }else {
                return ReturnMsg.msg(HttpServletResponse.SC_BAD_REQUEST, "上传失败");
            }
        }
        return ReturnMsg.msg(HttpServletResponse.SC_BAD_REQUEST, "格式错误");
    }
}
```

编写文件 SongServiceImpl.java，功能是验证上传歌曲的合法性，如果合法则将信息添加到数据库。在删除指定歌曲信息时，系统不但要删除数据库中的信息，而且要删除已经上传的音频文件和歌词文件。文件 SongServiceImpl.java 的具体实现代码如下。

```java
@Service("songService")
public class SongServiceImpl implements SongService{
    @Autowired
    private SongDao songDao;
    @Autowired
    private UserDao userDao;
    @Autowired
```

```java
    private TrendingRecDao trendingRecDao;
public List<Integer> getAllSongIdRecords() {
    return songDao.selectAllSongId();
}

public Song getSongById(int songId) {
    return songDao.selectSongById(songId);
}
public Song getSongByIdWithCollectionFlag(HttpServletRequest request, int songId) {
    //获取对应Id的歌曲
    Song song=songDao.selectSongById(songId);
    if(song==null) {
        return null;
    }
    //获取对应Id歌曲的流行度
    int trendingCoefficient=songDao.selectCoefficientById(songId);
    song.setTrendingCoefficient(trendingCoefficient);
    //获取用户的收藏列表
    List<Collection> collectionList=new ArrayList<Collection>();
    User user=userDao.selectByUser
            (Request.getUserFromHttpServletRequest(request));
    collectionList=trendingRecDao.getCollection(user);
    if(collectionList!=null) {
        for(Collection c:collectionList) {
            if(c.getSongId()==songId) {
                song.setWhetherCollected(true);
                break;
            }
        }
    }
    return song;
}

public void batchDeleteById(HttpServletRequest request,int[] songIds) {
    if(songIds==null) {
        return;
    }
    for(int id:songIds) {
        Song song=songDao.selectSongById(id);
        if(song!=null) {
            String realSongPath=request.getServletContext().
                            getRealPath(song.getSongAddress());
            File fileSong=new File(realSongPath);
            fileSong.delete();
            if(song.getLyricAddress()!=null) {
                String realLyricPath=request.getServletContext().
                            getRealPath(song.getLyricAddress());
```

```java
                    File fileLyric=new File(realLyricPath);
                    fileLyric.delete();
                }
            }
        }
        songDao.deleteByIds(songIds);
    }

    /**
     * 由于在前端文件上传的时候已经做过验证了，所以后端不再验证了
     * 这里默认前端的验证是可靠的
     */
    public boolean checkFormat(MultipartFile song, MultipartFile lyric) {

        return true;
    }

    public boolean addSong(HttpServletRequest request,MultipartFile song,
                        MultipartFile lyric) {
        String name=song.getOriginalFilename();
        //歌曲名称需去掉.mp3后缀
        String songName=name.substring(0, name.lastIndexOf("."));
        String songAddress="track/song/"+name;
        boolean isInsertSuccessful=false;
        int affectedRows=-1;
        //歌词文件是可选的
        if(lyric.isEmpty()) {
            affectedRows=songDao.insertOnlySong(new Song(songName,songAddress));
            //保存歌曲文件

saveFile(song,request.getServletContext().getRealPath(songAddress));
        }else {
            //这里的歌曲名称仍旧保留.lrc后缀
            String lyricName=lyric.getOriginalFilename();
            String lyricAddress="track/lyric/"+lyricName;
            affectedRows=songDao.insertSongWithLyric(new Song(songName,
                        songAddress,lyricName,lyricAddress));
            //保存歌曲文件
            saveFile(song,request.getServletContext().getRealPath(songAddress));
            //保存歌词文件
            saveFile(lyric,request.getServletContext().getRealPath(lyricAddress));
        }

        if(affectedRows>0) {
            isInsertSuccessful=true;
        }
```

```
        return isInsertSuccessful;
    }
    private void saveFile(MultipartFile multipartFile, String realFilePath) {
        try {
            InputStream inputStream=multipartFile.getInputStream();
            FileOutputStream fileOutputStream = new FileOutputStream(realFilePath);
            try {
                int b = 0;
                while ((b = inputStream.read()) != -1) {
                fileOutputStream.write(b);
            }
            }finally{
                inputStream.close();
                fileOutputStream.close();
            }

        } catch (IOException e) {
            throw new RuntimeException(e);
        }

    }
    public List<Song> getAllSongRecordsWithLyric() {
        return songDao.selectAllSongsWithLyric();
    }
}
```

8.7 排行榜模块

本系统提供了热歌排行榜功能，包括周榜排行和月榜排行，分别表示当周和当月播放量最高的歌曲推荐。本节将详细讲解排行榜模块的实现过程。

扫码看视频

8.7.1 获取数据库数据

编写文件 RankingPageDao.java，功能是获取数据库中的周排行榜和月排行榜数据，具体实现代码如下。

```
public interface RankingPageDao {
    /**
     * 获取最近一周排行榜列表
     * 如果没有，则返回 null
     */
    List<Song> selectRecentWeekRanking();

    /**
```

```
     * 获取最近一个月排行榜列表
     * 如果没有,则返回null
     */
    List<Song> selectRecentMonthRanking();
}
```

8.7.2 展示排行榜数据

编写文件RankingPageServiceImpl.java,功能是根据用户的选择展示对应的信息,即单击"周榜"链接显示周排行榜信息,单击"月榜"链接显示月排行榜信息。文件RankingPageServiceImpl.java的具体实现代码如下。

```
@Service("rankingPageService")
public class RankingPageServiceImpl implements RankingPageService{
    @Autowired
    private RankingPageDao rankingPageDao;
    @Autowired
    private UserDao userDao;
    @Autowired
    private TrendingRecDao trendingRecDao;

    public List<Song> getRankWithCollectionFlag(HttpServletRequest request, int
            mode) {
        List<Song> rankingPageList = new ArrayList<Song>();
        List<Collection> collectionList = new ArrayList<Collection>();
        User user = userDao.selectByUser(Request.getUserFromHttpServletRequest
                (request));
        collectionList = trendingRecDao.getCollection(user);
        if(mode==1) {
            rankingPageList=rankingPageDao.selectRecentWeekRanking();
        }else if(mode==2){
            rankingPageList=rankingPageDao.selectRecentMonthRanking();
        }else {
            //保留便于扩展
            rankingPageList=rankingPageDao.selectRecentMonthRanking();
        }

        // 在个性化列表中给已经被该用户收藏的歌曲加上标记
        if (collectionList != null && rankingPageList != null) {
            for (Collection c : collectionList) {
                for (Song t : rankingPageList) {
                    if (c.getSongId() == t.getSongId()) {
                        t.setWhetherCollected(true);
                    }
```

```
                }
            }
        }
        return rankingPageList;
    }
}
```

8.8 热门推荐模块

在本系统中，可以对播放量、用户评论量等数据进行综合分析，将得出的热门歌曲推荐给用户。本节将详细讲解热门推荐模块的实现过程。

扫码看视频

8.8.1 Controller 文件

编写文件 TrendingRecController.java，功能是将获取的热门推荐信息展示出来，具体实现代码如下。

```
public class TrendingRecController {
    @Autowired
    private TrendingRecService trendingRecService;

    @RequestMapping(value = "trendingRecFrameLoad.do",method = { RequestMethod.GET })
    public ModelAndView trendingRecFrameLoad(HttpServletRequest request) {
        ModelAndView modelAndView=new ModelAndView();
        modelAndView.setViewName("trendingRecFrame");
        List<Song> trendingSongList=
                trendingRecService.getSongWithCollectionFlag(request);
        modelAndView.addObject("trendingSongList",trendingSongList);
        modelAndView.addObject("test","Name");
        return modelAndView;
    }
}
```

8.8.2 获取数据库信息

编写文件 TrendingRecDaoImpl.xml，功能是获取数据库中的热门推荐信息，具体实现代码如下。

```
<?xml version="1.0" encoding="UTF-8" ?>
<!DOCTYPE mapper PUBLIC "-//mybatis.org//DTD Mapper 3.0//EN"
"http://mybatis.org/dtd/mybatis-3-mapper.dtd" >
<mapper namespace="com.haut.music.dao.TrendingRecDao">
```

```xml
<!-- 这里的trending为视图,返回歌曲的ID和其对应的流行系数
create view trending as
select songId,count(songId) as trendingCoefficient
from play group by songId limit 50;
-->
<select id="getTrendingSong"
    resultType="com.haut.music.model.Song">
    select * from trending,song
    where trending.songId=song.songId order by trendingCoefficient desc
</select>

<select id="getCollection" parameterType="com.haut.music.model.User"
    resultType="com.haut.music.model.Collection">
    select * from collection where userId=#{userId}
</select>
</mapper>
```

8.9 个性化推荐模块

本系统为每一位用户都提供了个性化推荐模块,使用深度学习框架 TensorFlow 和 Deeplearning4j 实现精准个性化推荐功能。本节将详细讲解个性化推荐模块的实现过程。

扫码看视频

8.9.1 展示个性化推荐信息

编写文件 PersonalRecController.java,功能是展示针对登录用户的个性化推荐信息,具体实现代码如下。

```java
public class PersonalRecController {
    @Autowired
    private PersonalRecService personalRecService;

    @RequestMapping(value = "personalizedRecFrameLoad.do",method = { RequestMethod.GET })
    public ModelAndView personalizedRecFrameLoad(HttpServletRequest request) {
        ModelAndView modelAndView=new ModelAndView();
        modelAndView.setViewName("personalizedRecFrame");
        List<Song> personalRecSongList=personalRecService.
                getPersonalDailyRecWithCollectionFlag(request);

        modelAndView.addObject("personalRecSongList",personalRecSongList);
        if(personalRecSongList==null) {
            modelAndView.addObject("oneDayOneWord","登录即享——遇见不一样的自己");
```

```
        }else {
            modelAndView.addObject("oneDayOneWord","更懂你的心");
        }
        return modelAndView;
    }
}
```

8.9.2 实现 ServiceImpl 类

编写文件 PersonalRecServiceImpl.java，实现个性化推荐的 ServiceImpl 类，具体流程如下。

(1) 编写方法 getPersonalDailyRecWithCollectionFlag()，功能是获取当前用户的个性化推荐列表。对应代码如下。

```
@Service("personalRecService")
public class PersonalRecServiceImpl implements PersonalRecService {
    @Autowired
    private PersonalRecDao personalRecDao;
    @Autowired
    private UserDao userDao;
    @Autowired
    private TrendingRecDao trendingRecDao;
    @Autowired
    private NewTrackOnShelfDao newTrackOnShelfDao;

    public List<Song> getPersonalDailyRecWithCollectionFlag(HttpServletRequest
            request) {
        List<Song> personalRecList = new ArrayList<Song>();
        List<Collection> collectionList = new ArrayList<Collection>();
        User user = userDao.selectByUser(Request.getUserFromHttpServletRequest
                (request));
        collectionList = trendingRecDao.getCollection(user);
        /* ============================================================ */
        personalRecList=selectPersonalRec(user);
        /* ============================================================ */
        // 在个性化列表中给已经被该用户收藏的歌曲加上标记
        if (collectionList != null && personalRecList != null) {
            for (Collection c : collectionList) {
                for (Song t : personalRecList) {
                    if (c.getSongId() == t.getSongId()) {
                        t.setWhetherCollected(true);
                    }
                }
            }
        }
```

```
        return personalRecList;
    }
```

(2) 编写方法 selectPersonalRec()，功能是每天早上 6 点更新个性化推荐列表，并从更新后的表中读取记录信息。对应代码如下。

```
/**
 * 这里采用两张表交替的方式来实现:
 * (1) 6 点之后就从另外一张表中读取记录
 * (2) 重新开始计算新的个性化推荐列表并存放于原来的表中
 * @param user
 * @return
 */
private List<Song> selectPersonalRec(User user) {
    if(user==null) return null;
    List<Song> personalRecList = new ArrayList<Song>();
    if(Static.isFromA) {
        personalRecList=personalRecDao.selectPersonalRecFromA(user);
    }else {
        personalRecList=personalRecDao.selectPersonalRecFromB(user);
    }
    return personalRecList;
}
```

(3) 编写方法 initializePersonalRecList()，功能是初始化个性推荐信息。对应代码如下。

```
public void initializePersonalRecList(HttpServletRequest request) {
    final User user = userDao.selectByUser
            (Request.getUserFromHttpServletRequest(request));
    List<Song> initialRecListA = new ArrayList<Song>();
    List<Song> initialRecListB = new ArrayList<Song>();
    //从新歌中随机获取 10 首，作为初始化列表
    initialRecListA=newTrackOnShelfDao.selecNewSong();
    for(int i=0;i<40;i++) {
        int len=initialRecListA.size();
        Random random=new Random();
        int index=random.nextInt(len);
        if(i<10) {
            initialRecListB.add(initialRecListA.get((index+1)%len));
        }
        initialRecListA.remove(index);
    }
    //批量插入
    if(Static.isFromA) {
        personalRecDao.insertListIntoRecA(initialRecListA,user.getUserId());
    }else {
        personalRecDao.insertListIntoRecB(initialRecListB,user.getUserId());
```

        }
    }

(4) 编写方法 updatePersonalRecIntoA()和 updatePersonalRecIntoB()，功能是分别更新歌曲推荐数据库表 a 和 b 的信息。对应代码如下。

```java
public void updatePersonalRecIntoB(Map<Integer, Integer[]> user2song) {
    user2song.forEach(new BiConsumer<Integer,Integer[]>(){
        public void accept(Integer userId, Integer[] recSongIds) {
            personalRecDao.deleteBByUserId(userId);
            //批量插入
            personalRecDao.insertArrayIntoRecB(recSongIds,userId);
        }
    });
}
public void updatePersonalRecIntoA(Map<Integer, Integer[]> user2song) {
    user2song.forEach(new BiConsumer<Integer,Integer[]>(){

        public void accept(Integer userId, Integer[] recSongIds) {
            personalRecDao.deleteAByUserId(userId);
            //批量插入
            personalRecDao.insertArrayIntoRecA(recSongIds,userId);

        }

    });
}
public void addHybridRecIntoA(Map<Integer, Integer[]> user2song) {
    user2song.forEach(new BiConsumer<Integer,Integer[]>(){
        public void accept(Integer userId, Integer[] recSongIds) {
            //批量插入
            personalRecDao.insertArrayIntoRecA(recSongIds,userId);
        }
    });
}
public void addHybridRecIntoB(Map<Integer, Integer[]> user2song) {
    user2song.forEach(new BiConsumer<Integer,Integer[]>(){

        public void accept(Integer userId, Integer[] recSongIds) {
            //批量插入
            personalRecDao.insertArrayIntoRecB(recSongIds,userId);
        }
    });
}
}
```

8.9.3 随机梯度下降算法

编写文件 SGD.java,功能是使用随机梯度下降算法实现个性化推荐,具体实现代码如下。

```java
public class SGD {
    public static void getW2WSourceDescent(Edge<String, String> edgeW2W, float[] ei, Set<String> bipartiteSetA,
            Map<String, float[]> wordEmbedding, Map<String, int[]> w2wNetwork, Set<String> wordSet, float[] descentW2WSource, float[] descentW2WDestination) {
        String sourceNode=edgeW2W.getSourceNode();
        String destinationNode=edgeW2W.getDestinateNode();
        float ej[]=wordEmbedding.get(destinationNode);
        int relatedArray[]=w2wNetwork.get(sourceNode);
        int len=relatedArray.length;
        //当前节点到对面的网络 B 的所有连边(公式的分母)
        float sumCurI2B=0;
        //当前节点到对面的网络 B 的所有连边乘 B 中对应的节点(公式中的分子)
        float sumCurI2BTimesNodeInB[]=new float[ei.length];

        //处理在双边网络 B 中的节点
        for(int i=0;i<len;i++) {
            String node=W2WNetwork.getW2WNode(i, wordSet);
            //获取对应节点的嵌入
            float[] eB=wordEmbedding.get(node);
            if(!bipartiteSetA.contains(node)) {
                //将双边网络 B 中的节点与当前节点的转置相乘
                float temp=(float) Math.exp(Operator.times(ei, eB));
                sumCurI2B+=temp;
                Operator.add(sumCurI2BTimesNodeInB,Operator.dotTimes(temp, eB));
            }
        }
        int weight=edgeW2W.getWeight();
        float[] descentSource=Operator.dotTimes(weight, Operator.dotMinus(ej,
                Operator.divide(sumCurI2BTimesNodeInB,sumCurI2B)));
        float[] curI2JTimesJ=Operator.dotTimes((float)
                        Math.exp(Operator.times(ei, ej)), ej);
        float[] descentDestination=Operator.dotTimes(weight,
                Operator.dotMinus(ei, Operator.divide(curI2JTimesJ,sumCurI2B)));

        Operator.assign(descentW2WSource, Operator.dotTimes(-1, descentSource));
        Operator.assign(descentW2WDestination, Operator.dotTimes(-1,
                    descentDestination));

    }
    public static void getW2DSourceDescent(Edge<String, Integer> edgeW2D, float[] ei,
```

```java
            Map<String, float[]> wordEmbedding, Map<Integer, float[]>
lyricEmbedding, Map<String, int[]> w2dNetwork, List<Song> engSongList,
            float[] descentW2DSource, float[] descentW2DDestination) {
        Integer documentNode=edgeW2D.getDestinateNode();
        float ej[]=lyricEmbedding.get(documentNode);
        //当前节点到对面的网络 B 的所有连边(公式的分母)
        float sumCurI2B=0;
        //当前节点到对面的网络 B 的所有连边乘 B 中对应的节点 (公式中的分子)
        float sumCurI2BTimesNodeInB[]=new float[ei.length];

        //处理在双边网络 B 中的节点
        for(Song song:engSongList) {
            //在文档子网络中歌曲 Id 表示的节点
            Integer node_j=song.getSongId();
            //获取对应节点的嵌入
            float[] eB=lyricEmbedding.get(node_j);
            //将双边网络 B 中的节点与当前节点的转置相乘
            float temp=(float) Math.exp(Operator.times(ei, eB));
            sumCurI2B+=temp;
            Operator.add(sumCurI2BTimesNodeInB,Operator.dotTimes(temp, eB));
        }
        int weight=edgeW2D.getWeight();
        float[] descentSource=Operator.dotTimes(weight, Operator.dotMinus(ej,
                    Operator.divide(sumCurI2BTimesNodeInB,sumCurI2B)));
        float[] curI2JTimesJ=Operator.dotTimes((float)
                    Math.exp(Operator.times(ei, ej)), ej);
        float[] descentDestination=Operator.dotTimes(weight,
            Operator.dotMinus(ei, Operator.divide(curI2JTimesJ,sumCurI2B)));
        Operator.assign(descentW2DSource, Operator.dotTimes(-1, descentSource));
        Operator.assign(descentW2DDestination, Operator.dotTimes(-1,
                    descentDestination));
    }
}
```

8.9.4　K 近邻分类算法

编写文件 UserKNN.java，功能是使用 K 近邻分类算法(KNN)实现个性化推荐，具体实现代码如下。

```java
public class UserKNN {
    /**
     * 获取用户的 k 个近邻用户
     * @param userIdList
     * 用户 Id 列表
     * @param user2songRatingMatrix
```

```java
 * 用户-歌曲"评分"矩阵
 * @param k
 * 参数k, k个邻居 */
public static Map<Integer, Integer[]> getKNN(List<Integer> userIdList, final
Map<Integer, float[]> user2songRatingMatrix, final int k) {
    // TODO Auto-generated method stub
    final Map<Integer,Integer[]> userKNNMatrix=new HashMap<Integer,Integer[]>();
    userIdList.forEach(new Consumer<Integer>() {

        public void accept(final Integer curUserId) {
            // TODO Auto-generated method stub
            Integer[] knnId=new Integer[k];
            //为用户建立一个最小堆来存放相似性最大的k个邻居
            final MininumHeap mininumHeap=new MininumHeap(k);
            //获取K Nearest Neighbors
            user2songRatingMatrix.forEach(new BiConsumer<Integer, float[]>() {

                public void accept(Integer otherUserId, float[] userRatingArray) {
                    // TODO Auto-generated method stub
                    //排除自己
                    if(otherUserId!=curUserId) {
                        //计算当前用户和其他用户的相似性
                        float similarity=Similarity.calculateSimilarity
(user2songRatingMatrix.get(curUserId),user2songRatingMatrix.get(otherUserId));
                        //放入堆中
                        mininumHeap.addElement(new TreeNode
                            (otherUserId,similarity));
                    }
                }
            });
            //从堆中获取相似性最大的k的邻居
            for(int i=0;i<k;i++) {
                knnId[i]=mininumHeap.getArray()[i].id;
            }
            userKNNMatrix.put(curUserId, knnId);
        }
    });
    return userKNNMatrix;
}
```

8.9.5 协同过滤算法

编写文件CollaborativeFiltering.java,功能是基于userKNN的协同过滤算法实现个性化推荐,具体实现代码如下。

```java
public class CollaborativeFiltering {
    /**
     * 基于最近邻用户产生协同过滤的推荐结果
     * @param userIdList
     * 用户 Id 列表
     * @param userKNNMatrix
     * 用户 KNN 矩阵
     * @param user2songRatingMatrix
     * 用户歌曲 "评分" 矩阵
     * @param songIdList
     * 歌曲 Id 列表
     * @param n
     * 推荐的前 n 首歌曲
     * @return
     * 用户歌曲推荐结果矩阵.userId,[recSongId1,recSongId2...recSongIdn]
     */
    public static Map<Integer, Integer[]> userKNNBasedCF(List<Integer> userIdList,
            final Map<Integer, Integer[]> userKNNMatrix, final Map<Integer,
                float[]> user2songRatingMatrix,
            final List<Integer> songIdList, final int n) {
        // TODO Auto-generated method stub
        final Map<Integer,Integer[]> user2songRecMatrix=new HashMap<Integer,
            Integer[]>();
        userIdList.forEach(new Consumer<Integer>() {

            public void accept(Integer curUserId) {
                // TODO Auto-generated method stub
                Integer[] knnIdArray=userKNNMatrix.get(curUserId);
                /**
                 * 对于每一首当前用户没有听过的歌曲
                 *     * 协同得分为:
                 * 其 k 个最近邻用户对该歌曲的 "评分" 的聚合
                 **/
                float[] curUserRatings=user2songRatingMatrix.get(curUserId);
                //为用户建立一个最小堆来存放评分最高的前 n 首歌曲
                MininumHeap mininumHeap=new MininumHeap(n);
                for(int i=0;i<curUserRatings.length;i++) {
                    //对于没有听过的歌曲
                    /**
                     * 这里需要注意的是，浮点数不能用==来比较
                     * 故这里用 curUserRatings[i]<0.01f 来表示
                       curUserRatings[i]==0f
                     */
                    if(curUserRatings[i]<0.01f) {
                        for(int knnIndex=0;knnIndex<knnIdArray.length;knnIndex++) {
                            int knnId=knnIdArray[knnIndex];
                            float[] knnUserRatings=user2songRatingMatrix.get(knnId);
```

```
                        curUserRatings[i]+=knnUserRatings[i];
                    }
                    //这里的聚合策略取均值
                    curUserRatings[i]/=knnIdArray.length;
                    int curSongId=songIdList.get(i);
                    //放入堆中
                    mininumHeap.addElement(new TreeNode
                        (curSongId,curUserRatings[i]));
                }
            }
            /**
             * 对该用户没有听过的歌曲，协同得分完成，选取 n 个得分最高的项目作为推荐
             */
            int trueNumber=n;
            //如果推荐的歌曲少于计划推荐的 n 首(处理歌曲很少的情况)
            if(mininumHeap.getCurHeapSize()<n) {
                trueNumber=mininumHeap.getCurHeapSize();
            }
            Integer[] curUserRecSongId=new Integer[trueNumber];
            for(int i=0;i<trueNumber;i++) {
                int recSongId=mininumHeap.getArray()[i].id;
                curUserRecSongId[i]=recSongId;
            }
            user2songRecMatrix.put(curUserId, curUserRecSongId);

        }
    });
    return user2songRecMatrix;
}
```

8.9.6 数据转换

编写文件 DataTranslate.java，实现数据转换功能，将用户相关信息和歌曲信息转换为矩阵，具体实现流程如下。

(1) 编写方法 getFrequencyMatrix()，构建用户频率矩阵来近似用户评分数据信息。对于某些系统而言，是不可能获取到用户对某些项目评分的，但是可以利用用户的行为习惯来反映用户的"评分"，比如一个用户常常收听某一首歌，那么我们可以推断该用户喜欢该歌曲的可能性很大。评分时，总分为 10 分，主动播放一次计 1 分，下载计 2 分，收藏计 5 分，如果超过 10 分，按 10 分计算。对应代码如下。

```
public static Map<Integer, float[]> getFrequencyMatrix(List<Integer>
    userIdList, final List<Integer> songIdList,
```

```java
                List<DownloadRecord> downloadList, List<PlayRecord> playList,
                    List<Collection> collectionList) {
            // TODO Auto-generated method stub
            final Map<Integer,float[]> user2songRatingMatrix=new HashMap<Integer,
                float[]>();
            final int songLen=songIdList.size();
            //获取用户-歌曲 下载映射
            final Map<Integer,Map<Integer,Set<Integer>>>
userId2songIdDownloadMap=getUserId2songIdRecordMap(downloadList,false);
            //获取用户-歌曲 收藏映射
            final Map<Integer, Map<Integer, Set<Integer>>>
userId2songIdCollectionMap=getUserId2songIdRecordMap(collectionList,false);
            //获取用户-歌曲-次数 播放映射
            final Map<Integer, Map<Integer, Set<Integer>>>
userId2songIdPlayMap=getUserId2songIdRecordMap(playList,true);
            userIdList.forEach(new Consumer<Integer>() {
                public void accept(Integer userId) {
                    // TODO Auto-generated method stub
                    float[] curUserRatingArray=new float[songLen];
                    int songIndex=0;
                    //处理每一首歌曲
                    for(Integer songId:songIdList) {
                        /**
                         * 处理下载，这里不考虑下载次数
                         */
                        if(userId2songIdDownloadMap.get(userId)!=null &&
userId2songIdDownloadMap.get(userId).get(SONG_ID_SET_KEY).contains(songId)) {
                            //当前用户下载过的歌曲
                            curUserRatingArray[songIndex]+=DOWNLOAD_SCORE;
                        }

                        /**
                         * 处理收藏，这里没有次数
                         */
                        if(userId2songIdCollectionMap.get(userId)!=null &&
userId2songIdCollectionMap.get(userId).get(SONG_ID_SET_KEY).contains(songId)) {
                            //当前用户收藏的歌曲
                            curUserRatingArray[songIndex]+=COLLECTION_SCORE;
                        }

                        /**
                         * 处理播放，考虑播放次数
                         */
                        if(userId2songIdPlayMap.get(userId)!=null &&
userId2songIdPlayMap.get(userId).get(SONG_ID_SET_KEY).contains(songId)) {
                            //当前用户播放过的歌曲
```

```
                          int count=userId2songIdPlayMap.get(userId).
                                get(songId).iterator().next();
                          curUserRatingArray[songIndex]+=PLAY_SCORE + count;
                      }

                      /**
                       * 处理最大得分，超过最大得分，记为最大得分
                       */
                      if(curUserRatingArray[songIndex]>MAX_SCORE) {
                          curUserRatingArray[songIndex]=MAX_SCORE;
                      }
                      //处理下一首歌
                      songIndex++;
                  }
                  //处理完一个用户
                  user2songRatingMatrix.put(userId, curUserRatingArray);
              }

          });
          return user2songRatingMatrix;
      }
```

(2) 编写方法 getUserId2songIdRecordMap()，构建用户 Id 跟歌曲 Id 的映射。对应代码如下。

```
      /**
       * 获取用户 Id - 歌曲 Id 的映射 Map
       * @param recordList
       * 包含 userId、songId 的记录列表
       * @param isCount
       * 是否需要计数。如果值为 true，则用 Integer[1]存放计数
       * @return
       * 两层 Map
       * 第一层 Map<Integer,Map> 每个 userId 拥有一个自己的 Map:
       * userId,userSetMap
       *
       * 第二层 Map<Integer,Set> 用户自己的 Map 里面存放两个数据:
       * (1)为每首歌曲计数 songId,CountSet;
       * (2)存放出现过的歌曲 songSetFlay,SongIdSet:
       */
      private static <T> Map<Integer, Map<Integer, Set<Integer>>>
      getUserId2songIdRecordMap(final List<T> recordList,final boolean isCount) {
          // TODO Auto-generated method stub
          final Map<Integer, Map<Integer, Set<Integer>>> userId2songIdRecordMap=
              new HashMap<Integer, Map<Integer, Set<Integer>>>();

          recordList.forEach(new Consumer<T>() {
```

```java
public void accept(T t) {
    // TODO Auto-generated method stub
    try {
        //利用反射和泛型获取不同类型表的相同属性
        //经常需要将特定对象转换成想要的 JSON 对象，为了实现通用性，用反射去
        //实现这个过程
        //获取当前对象的成员变量的类型
        //对成员变量重新设值
        Field userIdField=t.getClass().getDeclaredField("userId");
        Field songIdField=t.getClass().getDeclaredField("songId");
        userIdField.setAccessible(true);
        songIdField.setAccessible(true);
        int userId=userIdField.getInt(t);
        int songId=songIdField.getInt(t);
        //不需要计数
        if(!isCount) {
            //map 外层的 userId 已经存在
            if(userId2songIdRecordMap.containsKey(userId)) {
                //获取当前用户的记录集合 Map
                Map<Integer,Set<Integer>> curRecordSetMap=
                        userId2songIdRecordMap.get(userId);
                //将当前歌曲添加到当前用户的记录集合中
                curRecordSetMap.get(SONG_ID_SET_KEY).add(songId);
            }else {
                Map<Integer,Set<Integer>> curRecordSetMap=new
                        HashMap<Integer, Set<Integer>>();
                //创建记录歌曲 Id 的集合
                Set<Integer> curSongIdSet=new HashSet<Integer>();
                curSongIdSet.add(songId);
                curRecordSetMap.put(SONG_ID_SET_KEY, curSongIdSet);
                userId2songIdRecordMap.put(userId, curRecordSetMap);
            }
        }else {
            //map 外层的 userId 已经存在
            if(userId2songIdRecordMap.containsKey(userId)) {
                //获取当前用户的记录集合 Map
                Map<Integer,Set<Integer>> curRecordSetMap=
                        userId2songIdRecordMap.get(userId);
                //将当前歌曲添加到当前用户的记录集合中
                curRecordSetMap.get(SONG_ID_SET_KEY).add(songId);

                //计数
                count(songId,curRecordSetMap);
```

```java
                    }else {
                        Map<Integer,Set<Integer>> curRecordSetMap=new
                            HashMap<Integer, Set<Integer>>();
                        //创建记录歌曲 Id 的集合
                        Set<Integer> curSongIdSet=new HashSet<Integer>();
                        curSongIdSet.add(songId);
                        curRecordSetMap.put(SONG_ID_SET_KEY, curSongIdSet);
                        userId2songIdRecordMap.put(userId, curRecordSetMap);

                        //计数
                        count(songId,curRecordSetMap);

                    }
                }

            }catch (NoSuchFieldException e) {
                e.printStackTrace();
            } catch (IllegalArgumentException e) {
                e.printStackTrace();
            } catch (IllegalAccessException e) {
                e.printStackTrace();
            }
        }

        private void count(int songId, Map<Integer, Set<Integer>> curRecordSetMap) {
            /**
             * 计数,如果 Map<songId,count>已经存在,则直接计数+1
             */
            if(curRecordSetMap.containsKey(songId)) {
                //获取当前用户歌曲的计数集合(只有一个元素)
                Set<Integer> curCountSet=curRecordSetMap.get(songId);
                int cnt=curCountSet.iterator().next()+1;
                curCountSet.clear();
                curCountSet.add(cnt);
            }else {
                Set<Integer> curCountSet=new HashSet<Integer>();
                curCountSet.add(1);
                curRecordSetMap.put(songId, curCountSet);
            }
        }
    });
    return userId2songIdRecordMap;
}
```

8.10 项目测试

本项目运行后的主界面如图 8-17 所示,新用户注册页面如图 8-18 所示。

图 8-17 系统主界面

扫码看视频

图 8-18 新用户注册页面

排行榜页面如图 8-19 所示。个性化推荐页面如图 8-20 所示。添加歌曲页面如图 8-21 所示。

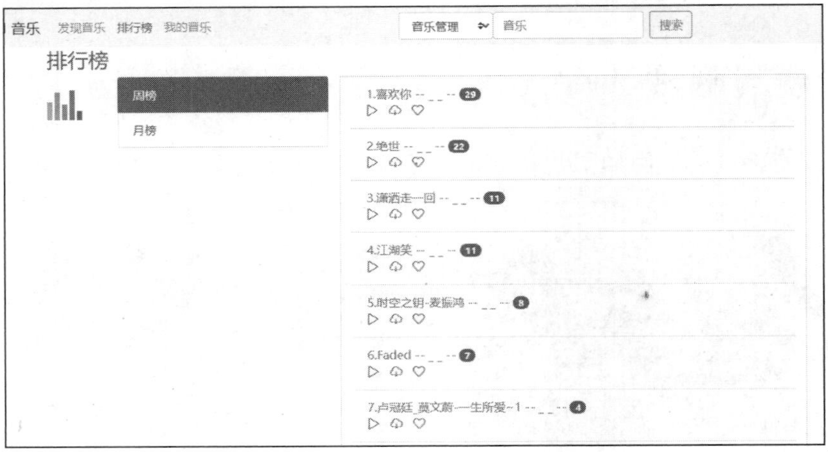

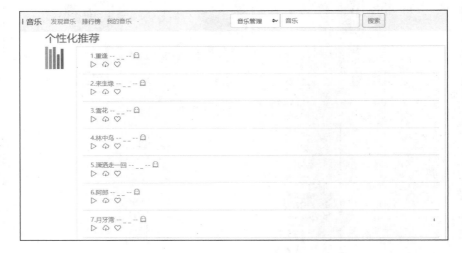

图 8-19　排行榜页面

图 8-20　个性化推荐页面

图 8-21　添加歌曲页面